William R. Park
CONSTRUCTION BIDDING FOR PROFIT

J. Stewart Stein
CONSTRUCTION GLOSSARY: AN ENCYCLOPEDIC REFERENCE AND MANUAL

James E. Clyde
CONSTRUCTION INSPECTION: A FIELD GUIDE TO PRACTICE

Harold J. Rosen and Philip M. Bennett
CONSTRUCTION MATERIALS EVALUATION AND SELECTION: A SYSTEMATIC APPROACH

C. R. Tumblin
CONSTRUCTION COST ESTIMATES

Harvey V. Debo and Leo Diamant
CONSTRUCTION SUPERINTENDENT'S JOB GUIDE

Oktay Ural, Editor
CONSTRUCTION OF LOWER-COST HOUSING

Robert M. Koerner and Joseph P. Welsh
CONSTRUCTION AND GEOTECHNICAL ENGINEERING USING SYNTHETIC FABRICS

CONSTRUCTION OF LOWER-COST HOUSING

CONSTRUCTION OF LOWER-COST HOUSING

OKTAY URAL

Professor & Director
International Institute for Housing & Building
Florida International University
Miami, Florida

A Wiley-Interscience Publication

JOHN WILEY & SONS
New York   Chichester   Brisbane   Toronto

TH
4812
C66

Copyright © 1980 by John Wiley & Sons, Inc.

All rights reserved. Published simultaneously in Canada.

Reproduction or translation of any part of this work beyond that permitted by Sections 107 or 108 of the 1976 United States Copyright Act without the permission of the copyright owner is unlawful. Requests for permission or further information should be addressed to the Permissions Department, John Wiley & Sons, Inc.

Library of Congress Cataloging in Publication Data

Main entry under title:
Construction of lower-cost housing.

(Wiley series of practical construction guides)
"A Wiley-Interscience publication."
Bibliography: p.
Includes index.

1. Dwellings.   2. Housing.   I. Ural, Oktay.
TH4812.C66   690'.8'1   79-19825
ISBN 0-471-89643-8

Printed in the United States of America

10 9 8 7 6 5 4 3 2 1

TO
NURSEL, CIGDEM, DERIN & SEVGI

Series Preface

The Wiley Series of Practical Construction Guides provides the working constructor with up-to-date information that can help to increase the job profit margin. These guidebooks, which are scaled mainly for practice, but include the necessary theory and design, should aid a construction contractor in approaching work problems with more knowledgeable confidence. The guides should be useful also to engineers, architects, planners, specification writers, project managers, superintendents, materials and equipment manufacturers, and, the source of all these callings, instructors and their students.

Construction in the United States alone will reach $250 billion a year in the early 1980s. In all nations, the business of building will continue to grow at a phenomenal rate, because the population proliferation demands new living, working, and recreational facilities. This construction will have to be more substantial, thus demanding a more professional performance from the contractor. Before science and technology had seriously affected the ideas, job plans, financing, and erection of structures, most contractors developed their know-how by field trial-and-error. Wheels, small and large, were constantly being reinvented in all sectors, because there was no interchange of knowledge. The current complexity of construction, even in more rural areas, has revealed a clear need for more proficient, professional methods and tools in both practice and learning.

Because construction is highly competitive, some practical technology is necessarily proprietary. But most practical day-to-day problems are common to the whole construction industry. These are the subjects for the Wiley Practical Construction Guides.

M. D. MORRIS, P.E.

Preface

This book is intended for a variety of readers. It is for undergraduate students who are interested in familiarizing themselves with a topic that concerns every family of the world; for builders who wish to learn more about new approaches to housing construction; for faculty members who plan to initiate new undergraduate courses in housing science; and for planners who are vitally concerned with reducing the cost of housing construction.

The book is conceived to consider every aspect of a housing activity. Lowering the cost of any unit depends on the type and quantity of materials used. Chapter 2 discusses the classical and rather new construction materials. It is extremely important that, if at all possible, indigenous materials be utilized for all construction. The technologies adapted affect the cost of building. Chapter 3 presents a thorough discussion of classical construction schemes with many illustrations. Since the industrialization of the building process is widely publicized as a powerful approach to help alleviate the housing shortage, the next four chapters discuss various aspects of this topic. Industrialization of the housing production sector can be realized through various steps. The primary industrialization can be the prefabrication of a few building components, such as doors, windows, and then of structural components, such as beams and columns. The second level of industrialization will be realized through the prefabrication of various panels, which can be the factory production of a complete unit such as a mobile home. The advantages and applicability of various levels of industrialization are clearly presented in Chapters 4, 5, 6, and 7. The immediate need for more new units is so pressing that it is necessary to plan the participation of the rural population in the construction of their shelters. This approach is known as "aided self-help," and Chapter 8 discusses this to great extent. Chapters 9 and 10 include information on housing problems in India and Venezuela. These concise chapters make clear many points about housing science.

Each chapter considers a distinct topic. When combined, the chapters interrelate and form the parts of an integrated systems approach.. It is with the aid of systems analysis that a meaningful result can be obtained. The knowledge acquired through experience and awareness of new technologies is the valuable asset in finding a complete solution to the world's housing problems.

I am indebted to Dan Morris for his continuous support on the conception and preparation of the manuscript. Also, I convey my sincere appreciation to professors Herbert Busching, Iraj Majzub, Ward Malisch, Norbert Schmidt, Douglas Wren, Fernando Tinoco, Edgar Wood, Hampapur Sreenath, and Adel Fareed for their contribution of excellent chapters. Needless to say, without their collaboration this book would not have been possible. My thanks especially to my wife Nursel who endured my long involvement in the project.

OKTAY URAL

Miami, Florida
November 1979

Contents

1. Lower-Cost Housing: Planning and Implementation   1
   Oktay Ural

2. Materials of Construction   15
   Ward Malisch and Norbert Schmidt

3. Classical Construction Methods for Low-Rise, Low-Cost Housing   35
   Douglas Wren

4. Semi-Industrialized Housing Production   123
   Herbert W. Busching

5. Industrialized Housing Production   139
   Edgar Wood

6. Building Systems Software   154
   Iraj Majzub

7. Systems Building Hardware   168
   Iraj Majzub

8. Aided Self-Help   230
   Adel Fareed

9. Housing Systems in India   245
   Hampapur Sreenath

10. Housing in Venezuela: Case Study   254
    Fernando Tinoco

    Bibliography   261

    Index   271

1

Lower-Cost Housing: Planning and Implementation

Oktay Ural

The word *housing* means much more today than it did a decade ago. The citizens of most countries are demanding not only a physical shelter, but also a home as a part of the total way of life. The quantitative demand is ever increasing as a result of uncontrolled population growth, migrations to urban centers, and the decay of existing housing units. The cumulative needs of many countries are already beyond easy corrective measures. A major concern of the governments is the potential social unrest due to the lack of decent shelters for the total population. Individuals who are awakening and becoming aware of the comforts created by the technological advancements are demanding more than a mere physical building as their home. It is this demand that forces the groups responsible for the planning to consider several new factors. To make a house a home is the key to success in handling the housing problems of the world. This can be achieved by emphasizing the human factors as well as the structural and planning aspects. The interdisciplinary nature of planning and producing a large number of housing units requires a team of experts who will be responsible for the success of the project.

The terms "low-cost housing" and "lower-cost housing" are widely used. Low-cost housing means housing for low-income population; it has different connotations for different parts of the world since the income levels are different. There are almost one billion people whose annual income is around $75.00, and they live in the developing countries. Low-cost housing that they can afford does not exist. Lower-cost housing, which has more scientific meaning, relates to lowering the actual cost of construction. This can be achieved through scientific and technological methods. Since the majority of the world's population falls into medium, low, and no-income groups, it is mandatory to stress lowering the cost of all construction. This is not a simple goal to achieve since centuries past have not been a sufficient enough time period to assure a successful result.

Planning

A large housing project, having a multitude of activities, needs to be planned from conception to the end (see Figure 1.1). This can be best achieved through the application of the concept of "systems approach," as exemplified in Figure 1.2. Housing technology and production have been comparatively slow in adapting scientific procedures for their expansion and improvement. A logical coordination among the stages for planning (including how health requirements will be met), designing, constructing, and maintaining a housing structure has been limited, if not completely missing. Experience has shown that a housing structure conceived in the light of its own functional value is not desirable since it cannot be a valuable part of a community. This particular approach is known as "systems building," which at one time was considered the only means to alleviate the housing needs of the countries. Because of the serious shortcomings of the systems building concept, which ignores many factors externally associated with the structure, it should not be used alone. However, it will be advantageous to incorporate it into a Systems Approach that considers all factors associated with the building. The scope of the studies related to the development of a large housing project goes far beyond the architectural and structural considerations. All decision makers were forced to observe the totality of the project. This awareness convinced everyone dealing with the planning of a housing project to recognize all factors that could influence its projected success. A process to interrelate these factors by well-defined logical connections was needed. It is this need which made the housing scientists turn to systems approach in order that they would produce successful and lasting results.

This approach will interrelate all the factors by assigning to them "weighing functions." Once this is done, an optimum result can be obtained for the total planning of the housing project. Even though this might not be accepted as an absolute optimum, it is one of the best results possible among several others. Systems approach is a flexible logical procedure adaptable to almost every developing country's housing problems. Its greatest asset is the inclusion of all human engineering input into use to

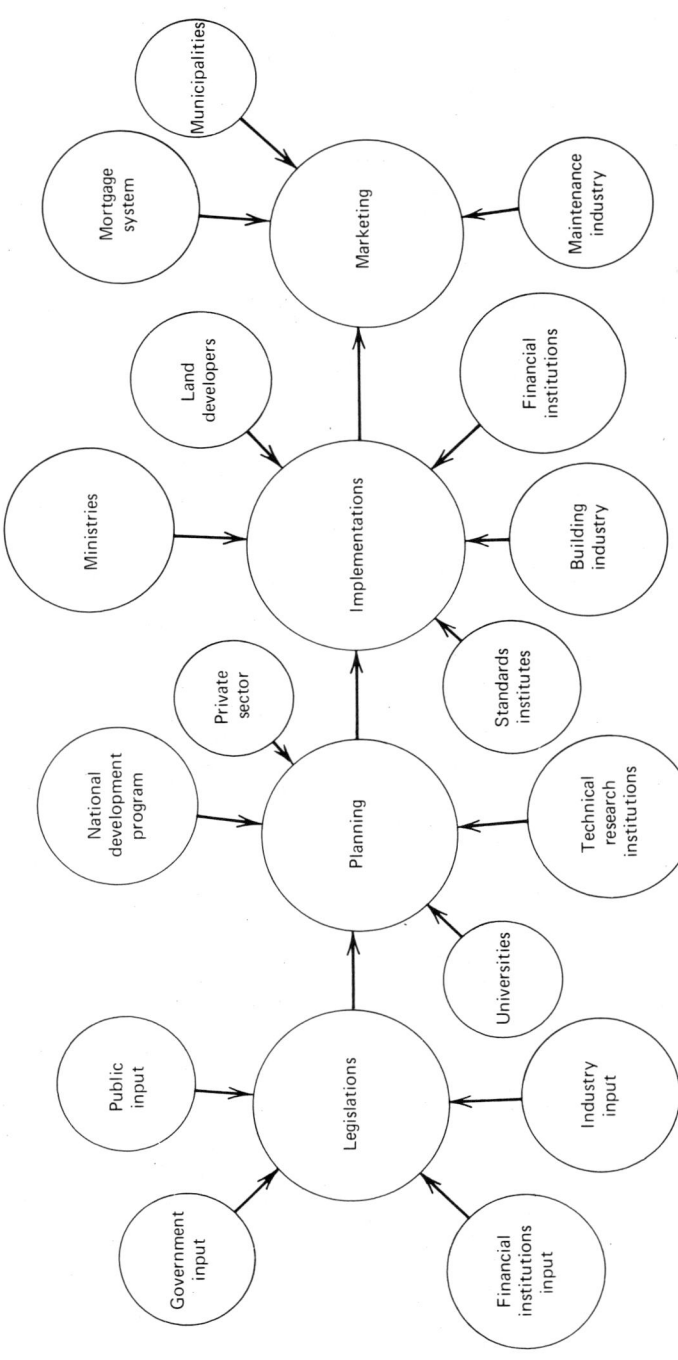

FIGURE 1.1. A flow diagram of a housing project activity.

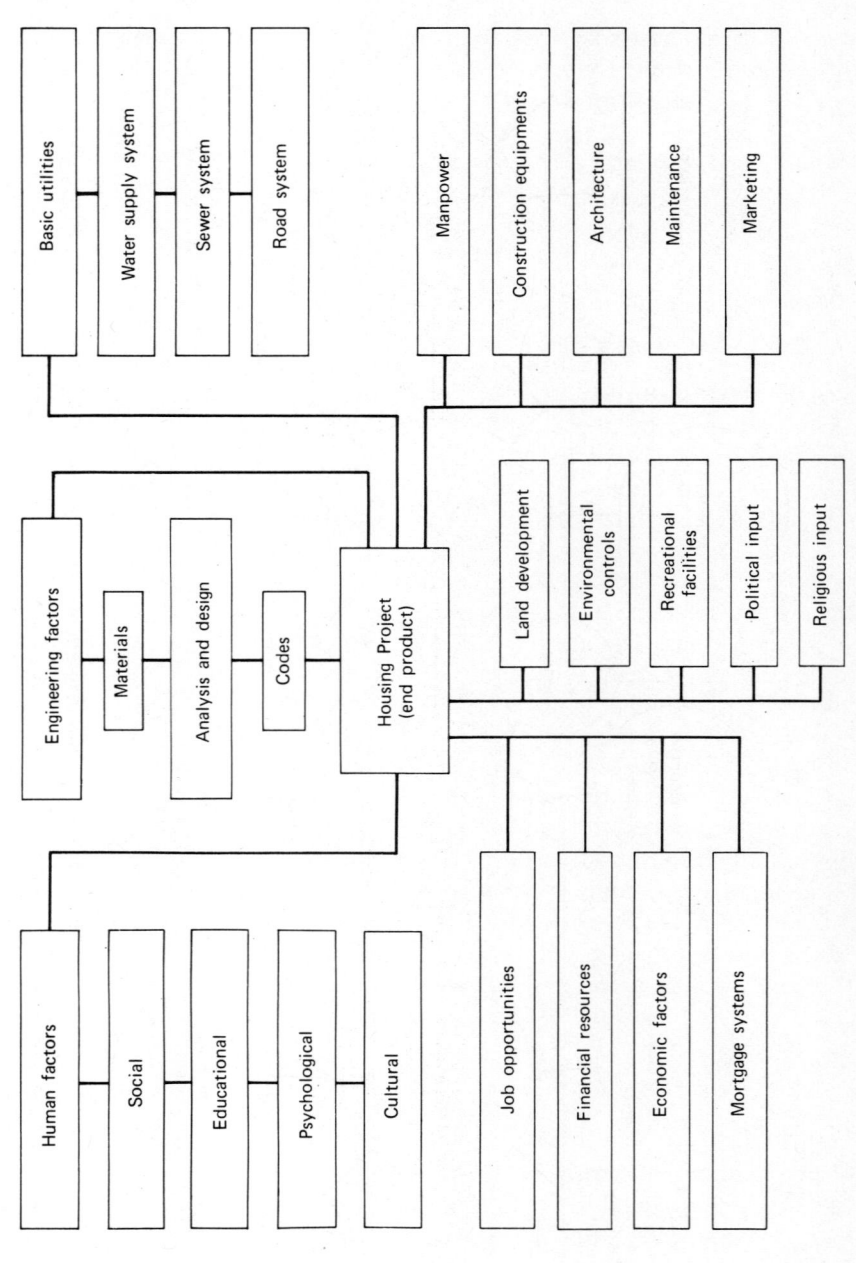

FIGURE 1.2. A system approach for a housing project.

reach one of the best solutions. Fundamentally, systems approach follows the reasoning of Rene Descartes's Discours sur la method. The four major logical steps to be performed can be concisely defined as follows:

1. Identify and recognize all relevant factors of the problem.
2. Assign realistic decision values to all defined factors.
3. Define the logic of the interconnections of all factors in order to be able to perform an analysis.
4. Optimize the result with respect to some well-defined dominant factors for best results.

Scientifically, step two of the systems approach is a complex one. It can be best performed by a competent interdisciplinary group aware that the factors to be considered are subjective in nature, such as the human factors. It is obvious that there exist constraints to define these by mathematical expressions which can be incorporated in a rigorous analysis.

The reference of relative evaluation is a "success contribution" to the total project. Each factor has an ultimate success contribution defined by the weighing function. In a simple study, constants can be used instead of weighing functions. The numerical assignments of the factors will be done with the aid of statistical information gathered during previous similar studies. The common sense, as well as professional information and experience, will be dominant assets in performing this task.

In the realm of this discussion, the Critical Path Method (CPM), when appropriate, needs to be incorporated in the planning process since its application will force the recognition of all relevant factors. CPM will define the time and duration of each activity. This is the accurate way to schedule manpower, equipment, materials. There are other project-control approaches that are basically the expansion of the Critical Path Method. One of these approaches, which has a worldwide accessibility, is ICES-PROJECT II. It is a computer-oriented effective method to enlighten and direct project engineers to accomplish their duties in the allocated time.

Urbanized Land

The procurement of proper land for housing construction is the initial activity. As the available land is becoming scarce, its cost is affecting adversely all housing projects. In urban areas, usually the land has the necessary infrastructures. Land cost is reaching as high as 50% of the purchase price of an apartment. Land speculations have become an economic and social issue in developing countries. The general trend is to force the governments to control the urban land prices and is rather a political decision to be reached by individual countries. One remedy to the situation is to decentralize the cities by encouraging the housing developments in the suburban regions which can be connected to urban centers by efficient mass transit systems. This will increase the available land for construction and help to improve the cities. Another planning idea is to develop satellite cities. An example of this is the Columbia City built between Baltimore and Washington, D.C.

The major need for decent low-cost housing is in the rural regions since the large portion of the population of the developing countries live in the countryside. The availability of land is not critical. What is critical is the lack of necessary infrastructures for a healthy life. Since the superstructures cannot be built unless the land is urbanized, the government must see that this is accomplished. Running water, sewers, and roads must be built. To keep the cost down, these have to be carefully planned and designed. The indigineous manpower should be used in the construction process. To control the land speculations in the rural areas, the government land can be leased to an extended period instead of being sold.

Implementation

To reduce the cost of a unit one has to reduce the costs of land, materials, manpower, and technical services or parts of them (see Figure 1.3). The reduction of land costs can be realized through the government's handling of the land ownership and land distribution. This can also be done through more effective use of land. One can use less frontage and high-rise approach in the high-density urban areas. Structural engineering know-how can provide us with a safe and economical design to construct high-rise buildings. The value of land in the cities began to affect the cost of homes in a different way. A decade ago, because of the reasonable cost of land, most contractors found it preferable to build apartment buildings only 4 or 5 stories high. Today, most of these structurally sound buildings are torn down to allow construction of 8- to 10- story apartment buildings. It is obvious that this renewal activity does not help the national economy. The unseen land effect on the construction industry has to be very carefully studied to have a healthy urban development program.

Many municipalities are using land-fill operations to claim new areas. Even though this can be profitable in marshy regions, it is wise to consider the potential difficulties to be encountered during the construction of the foundations. The provision of sites and services needs to be considered as an

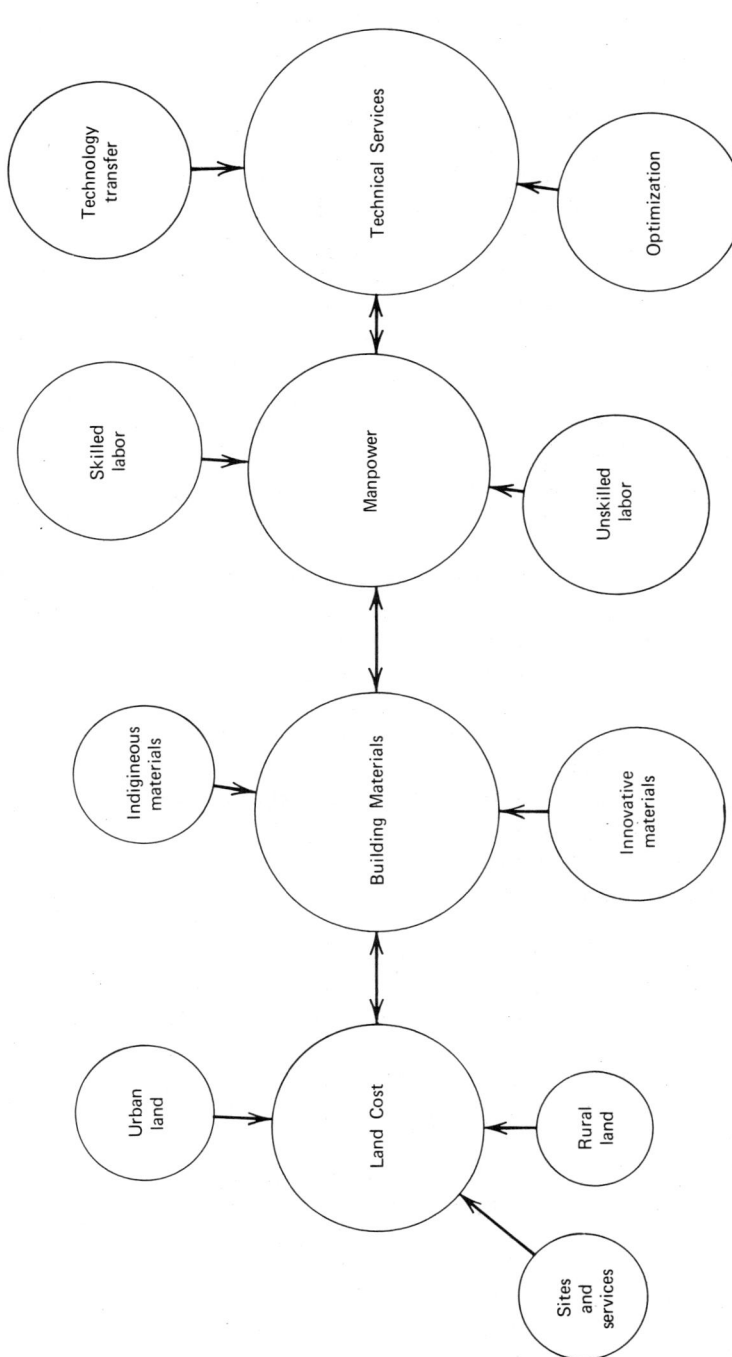

FIGURE 1.3. Major cost items in housing construction.

integral part of the land cost. As the cost of this service is decreased through the proper use of technology, the land cost will also be affected.

The building materials used in the construction of any housing unit forms more than half of the total cost. This is one item where a cost reduction can be realized if less materials are used through better design or cheaper materials are developed and used. It is true that in many countries buildings are overdesigned requiring more materials. There are also developing countries where parts of the construction materials are imported. Only the use of indigenous construction materials can help solve the housing problem of a country. Various types of bricks and bricketts are most widely used besides lumber. Cement and reinforcing steel are still not abundantly available, and when available they are relatively expensive. There are several innovative approaches such as the use of ferrocement and bamboo reinforcement. The scarcity of energy sources for heating and cooling is urging the material scientists to develop effective and cheap insulating materials. These could be natural or synthetic materials. Plastics and their derivatives are finding their way into the construction industry. However, they are not yet cost saving in the developing countries.

The labor force required for the construction of lower-cost housing should be carefully scheduled. In the developed countries the labor force in the building industry is expensive. Hence, these countries prefer to minimize the use of labor and replace it with appropriate machines. One reason for the initiation of the industrialization of the building sector is the rising labor cost. Prefabrication of many building components in central plants or on sites are becoming successful and cost saving in industrialized countries. The situation is different in the emerging countries where there is an abundance of unskilled and semiskilled labor force. In these countries, a labor intensive approach will decrease the cost of housing construction. Since time is money, the work force should be kept busy in a productive way. The preparation of the building site, foundations, and access roads can be achieved by unskilled labor force. Building blocks and bricks can be prepared again by unskilled laborers, avoiding capital-intensive machine operations. It is also true that emerging countries are beginning to adapt partial industrialization in their housing industries because of the massive need for new dwellings. These countries should plan to educate a work force to be able to function under the new system of construction.

The technical services will direct all operations of housing construction. Land development, site planning, infrastructures, structural analysis and design, and project control all need to be interrelated and scheduled according to the new methods. As these activities are controlled, the overall results will be optimized. Scientific and technological information are great assets to decrease the cost of construction. The technical services for a large housing project should be provided by an interdisciplinary technical team. Since there are many computer software, they should be used.

People

The ultimate goal of the activities of housing planning and construction is to provide an economical and acceptable dwelling for the people. This is a difficult task since the inclinations and desires of the people vary to a great extent; yet these must be determined and then used as data in the planning and decision processes. In the rural areas of emerging countries, almost without exception, single-story houses are preferred and require an addition for domestic animals and a small garden for vegetables. A high-density development in these regions will hardly be accepted. The style of architecture needs to be adaptable to the prevailing culture. The rural people do not object to adding a room or two as time goes by and their family needs demand more living space. It is obviously this fact that created the self-help concept. Every government should capitalize on this potential, which can serve two purposes: allow people to build as they like since this will serve their needs better and decrease the cost of housing units since the majority, if not all, of the labor will be provided by the individual owner. The owner will locate the cheapest possible materials for the expansion (see figures 1.4 and 1.5).

The economic development of countries occurs as they gradually become more industrialized than agricultural. The work force switches from jobs in the fields to those in the factories. Since the majority of the factories are located in or around urban centers, this change initiates a migration of people from rural areas to cities. These people find themselves to be little misfits in the new social and economic environment. There are no homes as their old ones and the available ones are high-rise apartment units. The purchase price or the monthly rent is beyond the economic limits of these people. Yet these people are determined to dwell in the cities and to work and enjoy the educational and social benefits of this new environment. It is this trend among rural people that created the squatters in and around the cities. The housing units in these regions are mostly built through self-help methods, and they are similar to the ones in the rural areas, except for higher density in the squatter areas. People who migrated feel more at home in these new communities, at least for a while. These transitory settlements are growing quickly, and their effects on the life of the cities are felt. It is a fact that the rural people who migrated to the urban centers are influencing the cities through their housing and communities. There are programs to replace the squatter housing by high-rise apartment buildings. There are

10   *Lower-Cost Housing: Planning and Implementation*

FIGURE 1.4.   A low-cost housing project, Lima, Peru.

such successful activities in Singapore; Hong Kong; San Juan, Puerto Rico; and Caracas, Venezuela. It seems Singapore has been a model city where high-density housing has been accepted and beautifully kept by the tenants. The people's contribution to the development of new decent communities is directly related to their cultural level—and even more to their formal educational level. In Caracas, Venezuela, the new tenants in high-rise apartments were asked to participate in night classes to learn how to take care of their new quarters. A similar educational program was initiated in Florida for the migrant workers who were living in fully furnished mobile homes.

The urban people began to move into high-rise apartment buildings as these buildings are replacing the old low rises. There are complaints from mothers since their children are enclosed in the building without easy and safe access to open spaces. It is obvious that people, if they have their choice, will prefer living in low-rise, separated units. In urban regions this is becoming a mere wish since the high land cost will prohibit any other consideration. See figures 1.6 and 1.7.

Government Action

Lowering the construction cost of housing units cannot be successfully accomplished unless the central or local governments make it their responsibility. This is necessary because a multitude of factors and contributors cannot be self-controlled. Already we have discussed several concepts and approaches; additionally, more detailed and technical concepts and approaches will follow, along with more detailed and technical information, in subsequent chapters. This is valuable information, and it will be helpful to the society when implemented. A nationwide action to build homes for low- and medium-income population can only be realized if the National Development Program of the particular country will allow it. According this program, the ministries of housing and construction, in every developing country, must lead the way by sponsoring the necessary legislation, which is mandatory to even initiate any serious program. I would like to stress this particular point, because the lack of enthusiasm from the governments is widely noticed. This can be attributed to the long-range nature of investments in national housing projects. The ministry acts as a catalyst for many activities to happen in harmony. The National Building Research Institutes in various countries are responsible for the scientific activities. They assess the adaptability of foreign technologies and building systems into the country. The evaluation of all foreign licenses to be used in the country can be correctly done in these institutes. If there is no such control, then it is possible to import too many different systems which might not all serve the intended results. It is advisable that open systems be prefer-

FIGURE 1.5. Squatter housing, Ankara, Turkey.

red over closed systems to have a wider applicability. The ministry can also encourage research on materials science not only to improve the existing building materials, but also to develop new innovative ones. It can develop a land control and distribution system for public low-cost housing construction.

To allow the low- and medium-income population of a country to own their homes depends on the availability of long-term loans. Since most emerging countries suffer high inflation rates, it is difficult to grant long-term loans. Only the central governments can establish an appropriate mortgage system. The management of this can be done through official financial institutions.

The construction of low-cost public housing needs a continuous and sufficient flow of funds. The government can provide and secure such support. As the urban population of each nation grows, the types of buildings will change and the living space allocated to each person will decrease. The governments can contribute to the decision on minimum-space allocation per family through a taxation system. This will encourage contractors to limit the living space of the apartment houses not to exceed a given value. Now, it is common to see homes with areas between 60 to 100 m². If properly designed, this will be sufficient for a healthy life of an average family.

FIGURE 1.6. A housing project near Caracas, Venezuela.

FIGURE 1.7. A low-cost housing project, Cairo, Egypt.

The government can also be influential in the incorporation of prefabrication in the building industry. The building industry is accustomed to operating according to the classical construction schemes and will not be open for rapid change. The ministries of construction can lead the way through their own initiative, especially in using the new building methods in official projects. The government is the most effective unit to initiate a national low-cost housing program and end it successfully.

Comments

The development of a housing project, in any country, to be successful, must follow various rigorous steps. This stems from the fact that today the need for new units reaches to 20 million per year because of the normal annual population growth. Assuming the demand for new shelters is exceeding the ability to provide, then the following steps must be taken as an integral part of any national development program:

1. Necessary legislation must be enacted to define the national housing policy. This policy will define and establish criterion that will be obeyed

by all disciplines and organizations responsible for the provision of all types of housing units.
2. Necessary legislation must be enacted to generate continuous and sufficient financial resources. This is mandatory to plan a successful national housing program. A powerful financial-resource system can be obtained by the establishment of a national housing program tax-like contribution. This can have a percentage base with respect to the income of each family. The monies can be collected in a special government institution or bank. This will secure the continuous flow of funds toward the realization of mass housing.
3. The municipalities should have a complete and long-range growth plan for their housing and housing-related construction. The building codes need to be strictly controlled without any exceptions. The municipalities, with the support and cooperation of the appropriate ministries, must prepare new housing development areas around the towns and cities. The infrastructures of these lands should be prepared before any superstructure is constructed.
4. The impact studies to define the influence of the new developments on the existing basic utilities must be finalized before the permit of the new development is approved. The neglect of this study will cause major problems in a short time and their corrections will be rather expensive, if not impossible.
5. The National Housing Policy should consider the prevailing land speculations and react to it by providing new urbanized land around the periphery of the existing urban centers.
6. The appropriate government organizations should support the implementation of proper industrialized housing production systems and use them in their own projects. The modular coordination should be nationally adapted. The industrialization needs to be implemented in parallel with the existing classical construction schemes.
7. The expansion of the research and evaluation of all available indigenous construction materials. The results of such activities will definitely introduce new materials and improve the use of other materials.
8. New technologies and management techniques need to be introduced to speed the construction activities in the nation. This will help to produce more and decrease the cost of eliminating the effects of inflation.
9. The decision process should consider all factors related to users' needs, costs, accessibility, job opportunity, density, environment, and financing.

These nine points emphasize the overall approach to the initiation of any large housing project.

2

Materials of Construction

Ward Malisch and Norbert Schmidt

The choice of materials for use in lower-cost housing units requires the consideration of several factors: structural properties, durability, insulating qualities, fire resistance, and unit weight. In addition, the material's ease of use, availability, and attractiveness are important.

An economic comparision of materials on a worldwide basis is complicated by variations in local labor costs, degree of industrial development, availability of raw materials, and construction methods in common use.

Therefore, no attempt is made to differentiate among the various materials on the basis of economy since this must be a local decision. The properties of each material are discussed and general advantages and disadvantages of these materials may be used as a basis for comparison.

The use of Portland cement concrete for lower-cost housing presumes that the necessary technological capabilities to produce cement and the raw materials are available. The raw materials that can be used include limestone, (or seashells to supply the lime) chalk, clay, shale, or silica sand to supply the silica. Additional materials such as iron ore may be necessary to supply the aluminum and iron. The manufacturing process must be carefully controlled to insure a uniformly high quality product.

Properties of Concrete

Concrete is unique since the manufacturing process is often performed on the building site. The quality of the concrete is therefore dependent on the skill of the worker to a greater degree than with some other building materials. (See Figure 2.1.)

FIGURE 2.1. Precast concrete unit is guided into position at Richard Allen Villa in San Antonio, Texas. The facility for low- and middle-income families has lightweight concrete walls and roof (Photo courtesy of the Portland Cement Association).

Concrete quality is a function of the quality of the cement-water paste, and paste quality depends on the ratio of water to cement used and the extent of "curing," that is, the time period during which concrete is kept moist at temperatures conducive to hydration. The longer the cement is kept moist, the more hydration takes place, which improves the quality of the concrete.

The water-cement ratio also affects quality. More water is used in mixing concrete than is needed for complete hydration; this water provides the necessary plasticity and workability. However, as the water content increases at a fixed cement content, strength and durability decrease.

Workability

Workability refers to the ease with which plastic concrete can be placed and compacted in a homogeneous mass. Workable concretes require a proper blend of aggregates and an adequate amount of paste to impart the desired plasticity. Higher water contents make concrete flow more readily, but reduce the strength and durability if cement content remains constant. For a given strength level (constant water-cement ratio), more fluid concretes generally require more cement and are thus more expensive. Consequently, the most economical mixture is the stiffest mixture that still has the required strength and workability.

Strength

Concrete compressive strengths usually range from 2500 to 6000 psi with higher water-cement ratios producing lower strengths. The longer the curing period the higher the strength will be, but the strength increases with curing time at a decreasing rate. At a fixed consistency, as required strength level increases, the cement content must be increased and thus the cost rises.

The tensile strength of concrete is considerably lower, usually ranging from 10 to 15% of the compressive strength. For this reason, concrete members subjected to tensile stresses are reinforced, usually with steel in the tension region.

Durability

Freezing and thawing of water-saturated concrete can result in deterioration such as surface scaling, spalling, and cracking. The most effective means for minimizing such damage is to add an air-entraining agent when mixing the concrete. The resulting concrete contains from 3 to 8% air in the form of minute dispersed bubbles that relieve pressures caused by freezing water. A low water-cement ratio is also beneficial since this results in stronger concrete that contains less freezable water and is less permeable.

Concrete may also be attacked by sulfates in the soil or groundwater. The use of sulfate-resisting cements, air entrainment, and pozzolanic admixtures have all been found to be effective in improving resistance to sulfate attack.

Other possible causes of durability distress in concrete include expansive reactions between alkalies in the cement and certain aggregates containing reactive silica and expansion of the cement paste as a result of the presence of excessive amounts of free lime or magnesia. The alkali-aggregate reaction can usually be controlled by using a low-alkali cement or a pozzolanic admixture in the concrete. Careful control in the cement manufacturing process can limit the amount of free lime and magnesia to tolerable levels.

Dimensional Stability

Concrete shrinks as it dries and when this shrinkage is restrained the resulting tensile stresses may cause cracking. The potential shrinkage can be minimized by use of the lowest possible mixing water content, but even high-quality, well-proportioned concrete mixtures will exhibit drying shrinkage. Joints are used to control the location of cracks, and reinforcing steel can be used to control crack widths. Concrete contracts when cooled and expands when heating. The coefficient of thermal expansion is roughly 6×10^6 in./in./°F, although it will vary with the aggregate used in the concrete and the moisture content. Cracking caused by thermal contraction can also be controlled by jointing and the use of reinforcement.

Load-induced instantaneous deformations in concrete can be calculated knowing the applied stress and the modulus of elasticity. The modulus of elasticity is estimated from the relationship $E = 33\gamma^{3/2}\sqrt{f_c}$ where E is the modulus of elasticity in pounds per square inch, γ is the unit weight in pounds per cubic foot and f_c is the compressive strength in pounds per square inch. Time-dependent or creep deformations may be as much as two to three times the instantaneous deformation.

Insulating Qualities

The thermal conductivity of normal-weight concrete usually ranges from 0.8 to 1.1 Btu ft/hr ft² °F. Concretes made using lightweight aggregates have considerably lower values, varying from 0.08 to 0.42 Btu ft/hr ft² °F.

Fire Resistance

Concrete is not flammable and generally possesses good fire resistance. However, strength and stiffness decrease significantly as the temperature is increased much above 200°F. The fire endurance of a concrete wall is primarily dependent on the wall thickness, quality of concrete, and types of aggregate and construction. Increasing wall thickness and greater cement content of the concrete both increase fire resistance. Lightweight aggregates are superior to siliceious aggregates with regard to fire resistance. The thickness of the cover over reinforcing steel, the bedding of masonry units, and whether the wall is solid or a hollow masonry unit all influence the fire resistance. Figure 2.2 shows an all-concrete home.

Unit-Weight–Normal-weight concretes have unit weights of approximately 140 to 150 pcf. Concretes made using lightweight aggregates may range in unit weight from 20 to 50 pcf for low-density insulating concrete to 85 to 120 pcf for structural concretes.

Wood Classification

Trees are divided into two classes: hardwoods, which are broad-leaved deciduous trees, and softwoods, which have narrow leaves and are evergreen. Softwoods are used primarily in structural applications such as framing and sheathing, while hardwoods are used mostly in flooring, wall and ceiling paneling, and trim. Wood may also be classified as heartwood or sapwood.

Wood Classification 19

FIGURE 2.2. An all-concrete home in Wasco, California. Wall sections are 4 ft cast-on-site panels, tilted into position and locked in place by perimeter apron-footing and cast roof (Photo courtesy of the Portland Cement Association).

Heartwood is the inner portion of the tree trunk consisting of inactive tissue and is surrounded by the sapwood, which contains living cells and takes part in the life processes of the tree. The natural decay resistance of heartwood is greater than that of sapwood if the wood is untreated.

Seasoning

Seasoning is defined as the process of removing moisture from lumber to improve its serviceability. When wood is dried below the fiber saturation point (about 25 to 35% moisture for most species) shrinkage occurs and there is an improvement in several properties. Thus seasoned lumber shrinks less and is lighter in weight. Seasoning may be accomplished by air or kiln drying.

Defects

The quality of a piece of lumber is affected by the presence of knots, grain irregularities, shakes (lengthwise grain separations between growth rings), checks (lengthwise grain separations across growth rings), warp, stain, and decay. Lumber is graded on the basis of appearance and/or allowable stresses by considering the basic properties for varying species as well as the extent and number of defects.

Properties

Wood differs from other housing materials in that less control can be exercised over the processes resulting in its internal structure. Properties vary from one species to another and will also be dependent on the defects present, seasoning, and preservative treatments used.

Strength

The strength values of wood free from defects are determined from laboratory tests on small specimens. These values vary depending on the direction of loading with respect to fiber direction, seasoning, and load duration. The compressive strength for small specimens of green wood tested parallel to the grain may range from 2300 to 5000 psi, with some dry woods having strengths in excess of 9000 psi. Flexural strength for small green wood specimens can vary from about 5,000 to 11,000 psi, with dry woods once again testing higher. The proportional limit for green wood tested perpendicular to the grain is on the order of 200 to 800 psi.

The increases in strength caused by drying are most pronounced in small specimens. Because of an increased probability of seasoning defects with an increase in size, the bending strength differences between seasoned and unseasoned lumber over 5 in. thick are very small. For this reason, allowable working stresses in members greater than 5 in. thick are not increased for continuously dry service.

Durability

Deterioration of wood may be caused by decay, marine borers, or insects. Decay results from the action of wood-destroying fungi whose growth and spread depend on moisture being present. Thus wood kept at a moisture content below the fiber saturation point generally will not decay. If no decay-preventing treatment is used, heartwood will resist decay better than sapwood, and some species—redwood, for example—have a high natural resistance to decay.

For optimum resistance to decay and borer or insect attack, wood can be treated with preservatives. Creosote is very effective, but its objectionable odor prohibits use in housing other than for pilings. Waterborne preservatives such as zinc chloride or organic materials such a pentachlorophenol can be successfully used in sills, joists, or other structural members that may be wet during service.

Pressure treatment is much more effective than dipping or painting the wood with a preservative, especially in warm and humid regions. However, special equipment is needed to obtain the necessary retention of the preservative.

Dimensional Stability

Wood shrinks when it dries below the fiber saturation point. If it is seasoned and installed at the equilibrium moisture content corresponding to service conditions, shrinkage is minimized. When larger unseasoned timbers are used it is necessary to reduce allowable loads at fastenings and sometimes to provide for the movement that will occur.

The modulus of elasticity for wood is relatively low compared to other building materials. It seldom exceeds 2×10^6 psi and this means that limiting deflections are an important consideration in the design of wood members.

Insulating Qualities

Solid wood is an excellent insulator with an average thermal conductivity for coniferous species of about 0.07 Btu ft/hr ft^2 °F. Wood frame walls with insulation in the space between exterior sheathing and interior plaster lath provide excellent insulating qualities.

Fire Resistance

The fire resistance of untreated wood is poor for members with small cross sections. When such untreated wood is subjected to temperatures of 300°F, flammable gases are produced and the wood begins to char. Small pieces burn quickly, but larger members retain their strength for some time because of the low charring rate and low thermal conductivity. For this reason, large timbers may perform better than metals under fire conditions.

For small frame construction, fire resistance can be imparted to wood by impregnating it with substantial quantities of suitable chemicals. Some degree of fire resistance can also be provided by fire resistant coatings.

Unit Weight

Wood is relatively light construction material with a unit weight ranging from 20 to 50 pcf. The unit weight depends on species, growth rate, moisture content, and whether the wood is treated for decay or fire resistance. Generally, the strength increases with increasing unit weight.

Wood Composites

Construction with solid timber is limited by the size and shape of the timbers available. Glued-laminated timbers can be used to overcome this disadvantage and also to provide members with more nearly uniform strength.

Plywood is also a glued-laminated wood product consisting of several thin layers of wood glued together with the grain of adjoining piles at a 90° angle to each other. The strength properties along the length and width of a panel are nearly equalized, permitting use in sheeting applications. The major disadvantage of plywood and glued-laminated construction is increased production costs.

Metals

Steel

Most of the steel used for construction is low- to medium-carbon steel that is tough, strong, and easy to work. An alloy steel is one in which various elements such as manganese, silicon, aluminum, titanium, or molybdenum are alloyed with carbon steel to produce properties unobtainable in carbon steel.

Structural steel is usually a medium-carbon steel consisting of hot-rolled sections, shapes, and plates. However, alloy, high-strength, or corrosion-resistant steels may also be used for the thinner gages or where the surface finish, mechanical properties, or closer tolerances that result from cold-reducing are desired.

Properties

The most commonly used structural steel (ASTM A 36) has a minimum yield point of 36,000 psi and a minimum tensile strength of 58,000 to 80,000 psi. There are several high-strength steels that have a greater structural abilities, with minimum yield points as high as 90,000 to 100,000 psi and minimum tensile strengths ranging from 105,000 to 135,000 psi.

Both rotproof and termiteproof, steel is not adversely affected by freezing and thawing cycles. Corrosion is the most common form of durability distress in exposed steels, but recently developed high-strength, low-alloy steels have a very high resistance to corrosion as well as higher strengths. Paints and metalic or chemical coatings can also provide corrosion protection.

Dimensional stability of steel is good, with a typical linear coefficient of expansion being a 6.5×10^{-6} in./in./°F at temperatures below 100°F. The modulus of elasticity of structural steel is about 29×10^6 psi.

Steel is a poor insulating material. The thermal conductivity is approximately 26 Btu ft/hr ft^2 °F. Although steel is a noncombustible material, its strength and stiffness decrease markedly at elevated temperatures. Consequently, for improved fire resistance, structural steel has been protected with many materials such as brick, stone, concrete, gypsum, mineral fibers, and fire resistant plasters.

Steel is also a relatively heavy building material with a unit weight on the order of 490 pcf.

Light-Gage Steel

Light-gage steel may be formed in three ways: from flat-rolled carbon steel; from sheet or strip steel that is cold-formed in roll-forming machines, press brakes, or bending brakes; or from sheet steel shaped into long, ribbed roof decking. Open-web steel joists are lightweight trusses composed of relatively small bars, bar-size angles and shapes formed from flat-rolled materials. While the components are not necessarily light-gage steel, open-web steel joists are frequently used in lightweight steel construction.

Many different kinds of lightweight structural components and building systems that use them have been developed. Combinations of different construction materials such as masonry-bearing walls and open-web steel joists are possible as well as conventional skeleton framing. Field connections can be made by bolting or using self-tapping screws or special fasteners.

Reinforcing Steel

Reinforcing steel for concrete is usually in the form of round, deformed bars or wire mesh. It comes in several grades with a minimum yield point ranging from 40,000 to 75,000 psi and minimum tensile strengths from 70,000 to 100,000 psi.

Aluminum Alloys

Aluminum alloys may be used for housing in the form of structural shapes, siding, sheet or corrugated sheet as well as for windows, doors, and other miscellaneous uses. A variety of mechanical, chemical, and coated finishes are available.

Properties

The yield strength of aluminum alloys can range up to 70,000 psi with an ultimate strength as high as 80,000 psi. They are highly corrosion resistant under most atmospheric conditions, but some will react with dissimilar metals, wet concrete, mortar, or plaster, and certain wood preservatives and must be insulated from contact with these materials.

The modulus of elasticity is considerably lower than that of structural steel, about 10.5×10^6 psi and the coefficient of linear expansion is higher also at 12.8×10^{-6} in./in./°F.

Thermal conductivity is very high, on the order of 128 Btu ft/hr ft² °F, and fire resistance is low unless the aluminum is coated with fire resistant materials. The unit weight is approximately 170 pcf.

Sheet Products

Corrugated sheet can be used for siding or roofing and has excellent weathering properties. It is strong, does not stain adjoining materials, and can be lapped to avoid joint problems and form a weather seal. Aluminum siding can be produced in many designs and textures with factory-applied coatings of sprayed-on, baked-on, or procelain enamel. Horizontal siding is produced with fiberboard laminated to the back for increased insulation and maximum ease of installation.

Sandwich panels may consist of a cellular core of aluminum or other insulating material with a skin of corrugated or sheet aluminum bonded to both sides. These panels may be used as curtain walls, supporting walls, floors, roofs, and partitions. They can be mass produced and combine excellent durability with good insulation qualities.

Other Materials

Masonry

Masonry refers to all types of building materials that consist of units held together with mortar—for example, stonework, brickwork, and concrete block construction. The properties of masonry work depend on the methods used to bond, reinforce, and tie the units into a whole.

Stone

Common stones used for building may be stratified or unstratified rocks of igneous, sedimentary, or metamorphic origin. Limestone, granite, sandstone, slate, and marble are common, but other types have also been used in building.

The compressive strength of commonly used building stone varies from 2,500 to 30,000 psi, but the compressive strength of stone masonry probably depends more on the strength of the mortar than on the strength of the stone itself. Flexural strength may be important if the stone is used for structural units such as lintels, where bending stresses are produced.

The durability of stone depends on the composition, structure, and grain size of the material. Freezing and thawing or leaching may damage the more porous stones, and acid such as sulfur dioxide may attack calcareous stones with a calcareous cementing material.

Dimensional stability of stones will vary also, depending on the thermal coefficient of expansion, which may range from 0.6 to 6.9 $\times$ 10^{-6} in./in./°F. Fire resistance of stone is also affected by the thermal coefficient of expansion and the composition. Heating and rapid cooling by water can cause cracking and spalling, while calcination and disintegration of limestone begins at about 1100 °F. Thermal conductivity of building stones is similar to that of concrete, but may be lower depending on the type of stone.

The unit weight of stones may range from 130 to 190 pcf.

Brick-building bricks are masonry units composed of inorganic nonmetallic materials hardened by heat or chemical action. They may be classified as adobe, made of sun-dried clays or earth and a binder; kiln-burned, composed of clays or shales to which other materials may have been added and fired to hardness; sand-lime, mixtures of sand and lime hardened under steam pressure and heat; or concrete, composed of Portland cement and aggregates. Bricks are available in a wide variety of types and sizes.

Most bricks will have compressive strengths from 1250 to 9000 psi with hard-burned bricks having higher compressive strengths. The modulus of elasticity may vary from 1.4 to 5.0 $\times$ 10^6 psi. Durability depends on the grade of the brick; common brick may be manufactured to resist exposure conditions ranging from heavy rain, snow, or continual freezing to minimal moisture and freezing.

Fire resistance of brick masonry walls is generally good, but wall thickness, type of wall, and whether combustible materials are framed in the wall thickness will all affect the fire-resistance rating. The thermal conductivity of brick masonry walls depends on brick density, wall thickness, and the type of wall. An average thermal conductivity of brick masonry walls depends on brick density, wall thickness, and the type of wall. An average thermal conductivity for clay brick is about 0.4 Btu ft/hr ft^2 °F. The unit weight of burned brick ranges from 105 to 140 pcf.

Concrete block-hollow or solid masonry units of Portland cement and sand, gravel or other aggregates are termed "concrete block." Block can be made with normal-weight or lightweight aggregate and in many sizes and shapes.

Compressive strength of concrete block is calculated based on the average gross area. Minimum compressive strengths may range from 1800 psi for solid load-bearing units to 350 psi for hollow nonload-bearing units. Durability is related to water absorption, and a maximum absorption of 15 lb/ft³ of concrete is frequently specified. A waterproof coating is recommended when block is used on exterior walls.

Dimensional stability is similar to that of concrete. Insulating and fire resistance values depend on the type of aggregate used, with lightweight aggregates producing blocks having lower coefficients of heat transmission and better fire-resistance ratings.

The unit weight also depends on the aggregate type used. Block made with sand and gravel or crushed stone weighs from 40 to 45 lb/8 × 8 × 16 in. unit, while blocks made with lightweight aggregate weigh from 25 to 35 lb per unit.

Mortar

Mortar for masonry units consists of cementitious materials (usually Portland cement and/or hydrated lime), sand, and water. Properties of mortars vary greatly, being dependent on the properties of the cementing agents and sand and on their relative proportions. Workability, water retention, strength, and dimensional stability are all important. Workability is controlled by the character of the cementitious material and the amount of sand, while strength is affected primarily by the amount of cement in the matrix. See Figure 2.3.

Plastics

Plastics are synthetic organic high polymers, all of which are plastic at some stage in their manufacture. Thermoplastics can be softened by heating and hardened by cooling any number of times while thermosetting plastics are converted by heating to a form that cannot be remelted. The essential ingredient of all plastics is a resin, but plasticizers, fillers, and colorants are added to achieve certain desired properties, facilitate processing, or reduce the cost.

Plastics are available with such a wide variety of physical and chemical properties that a full description is beyond the scope of this chapter. Compressive strengths may range from 1,000 to 70,000 psi and tensile strengths from 500 to 50,000 psi. Plastics will not corrode and are not attacked by insects or decay, but some may change color and/or craze under outdoor exposure conditions. The thermal coefficient of expansion is high and stiffness is relatively low. Thermal conductivity and unit weight are low, especially for the cellular plastics such as polyurethane and polystyrene. Most plastics will burn, but will not support combustion; thermoplastics will soften at elevated temperatures and thermosetting plastics cannot resist continuous heat at temperatures more than 550°F.

Other Materials 27

FIGURE 2.3. Econo-Line concrete masonry home built in Rocky Mount, North Carolina. The exterior walls are made with lightweight concrete block and the floor is a 4-in. thick concrete slab containing flat-sheet wire mesh (Photo courtesy of the Portland Cement Association).

The compounds to be converted into manufactured products are supplied in powder or liquid form. The molding powders are heated to form viscous liquids during processing. Thermosetting materials are commonly formed by either placing molding powder in heated dies and compressing under heat into the final shape or by forcing heat-softened material into a heated die. Thermoplastics may be formed by pouring into a mold and cooling, by injection molding, or by an extrusion process in which a viscous liquid thermoplastic is forced through an orifice shaped to yield the desired cross section. High- or low-pressure laminates are formed by impregnating sheets of cloth, paper, glass fiber, or other fabrics with liquid thermosetting plastics, stacking them in layers, and curing by heat and pressure. Plastic foams may be made by introducing air or some other gas into liquid thermosetting plastics, stacking them in layers, and curing by heat and pressure. Plastic foams may be made by introducing air or some other gas into liquid thermosetting plastic and solidifying by heating. Prefoamed or foamed-in-place plastics are available.

Indigenous Materials

In developing countries, many of the previously mentioned materials may be unavailable or of such high cost that housing construction using large quantities of cement, wood, or metals is not feasible. In many such countries, labor costs are low and labor-intensive housing construction methods using locally available materials offer the best solution to the lower-cost housing problem.

Indigenous materials may include soil, stone, lime, fly ash, coal ash, and some species of wood or plant life. Soil materials may be used in rammed-earth walls with or without bamboo reinforcement or in producing soil-cement and adobe blocks for masonry construction. Relatively small amounts of cement can be used to stabilize the soil or blocks, and this cement can be used to stabilize the soil or blocks and this method of manufacture does not require highly skilled labor.

Pozzolanic materials such as fly ash or volcanic ashes can be used as a cement replacement in concrete with up to 40% replacement being possible. Use of lime as an additive to the cement-pozzolan concrete generally improves the structural properties.

Bamboo has been used instead of steel as tension reinforcement in concrete, and successful results have been reported for composite reinforced concrete floor slabs using either semicorrugated or corrugated asbestos sheets in the tension zone to reduce steel costs.

The advantages of using lower-cost indigenous materials must be carefully studied, however, since the strength, durability, and overall quality of construction are generally low and maintenance costs may be high.

Soil

Definition

Soil has been defined as a natural aggregate of mineral particles, with or without organic admixture, that can be separated by gentle mechanical means such as agitation in water. One definition suggests that soil is the material with an unconfined compressive stress less than 200 (1378 kPa) as differentiated from rock, with a strength greater than 200 psi.

Classification

Soil has been classified by a multitude of classification systems: geologic, pedologic, engineering, and so on. Geologically, soil is either residual (formed in place from the parent rock) or transported. Agents of transportation identify glacial, fluvial (stream borne), lacustrine (lake), or aeolian (windborne) deposits. Generally a geologic description may also include a description of grain size or plasticity: gravel, sand, silt, or clay. Many informal descriptors are commonly used to classify geologic deposits, and no really formal system is recognized.

Pedologic soil classification is based on the five soil forming factors: climate, topography, vegetation, parent material, and time. This system is mainly applicable to the upper-weathered layers of soil that support vegetation. Although such classification systems were not designed to aid engineering, several factors are pertinent to both pedologic and engineering needs. For example, predominate grain size and permeability are inherent in a pedologic system. The fact that large areas of the world have been mapped pedologically allows a soils engineer familiar with the system to make valuable engineering inferences where such maps are current.

Engineering classification is based on the texture of the soil. The primary distinction is between coarse-grained or granular soils and fine-grained soils. Granular soils are further classified by predominate grain size to gravels or sands. Fine-grained soils are subdivided on the basis of plasticity, often by performing Atterberg Limit tests (ASTM Designations D423 and D424), as silts, which are nonplastic or slightly plastic, and clays, which are plastic.

Gravels or sands, when clean (very few fines) are further classified as well as graded or poorly graded on the basis of a mechanical (grain size) analysis. When the fines content is appreciable, the subclassification is based on the nature of the fines, either silty or clayey.

The presence of organic matter adds to the classification. Soils that are predominantly organic, rather than mineral, are peats and mucks. The Canadians have developed a separate classification system for peats and mucks.

The standard engineering classifications such as the Unified (ASTM Designation D2487) or the AASHTO systems do not consider the structural and moisture properties unique to the soil in its natural environment. These systems therefore have a limited usefulness to soils insitu, but are excellent for classifying the potential of a soil to be compacted in an embankment.

Properties

GENERAL   A number of physical properties are important to the engineering behavior of a soil: hydraulic properties, stress-strain or load-deformation properties, and shear strength, as well as the permanence of these properties. The shrink-swell behavior of soil is also of great engineering significance. There are many other properties that under special circumstances may be important: acidity or alkalinity and corrosive potential, erosion potential, or dispersiveness, thermal properties, frost action, capillarity, and toxicity, to name a few.

HYDRAULIC Soils can transmit or retard the flow of water through soil pores to greatly varying degrees and likewise water pressures can be transmitted. The permeability of a soil, its ability to transmit flow under a unit hydraulic gradient, is pertinent to its use as a drainage layer, a water-retarding core of an enbankment or simply in the analysis of a natural deposit when construction dewatering is required. Permeability varies from $k = 10^2$ cm/sec in openwork gravel to $k = 10^{-9}$ cm/sec in a sodium montmorillonite dense clay. This is a 100-billion-fold variation. Permeability depends on pore sizes and the continuity and orientation of pores of the soil. It is therefore a function of grain size, gradation and the void ratio (ratio of volume of voids to that of solids) of coarse grained soils and silts. In clays, the mineralogy, chemistry, and particle orientation are also of significance. Laboratory permeability tests may be performed on soils, with attempts made to control natural density and natural structure when possible. Results are often misleading because of insitu stratification and secondary jointing, which cannot be duplicated in the laboratory. For materials of higher permeability, the field pumping test, when properly conducted and interpreted, can produce meaningful permeability measurements.

The stress in a soil mass is shared by that stress carried by the intergranular or effective stress and is the total force carried by all the contacts divided by the total area of the soil mass. The contact area is very small; so the stress across any given contact may be quite large, but the effective stress is much less. The effective stress is effective in mobilizing the shear strength and compressibility of the soil. The pore pressures play no part in this mobilization and are often referred to as neutral stresses. When total stress is constant and pore pressures increase, the effective stress therefore decreases and strength decreases. A dissipation of pore pressure will increase strength and increase the compressive deformation of the soil. This principle of effective stress, introduced by Karl Terzaghi (1967) one of the reknown professor of soil mechanics, was the cornerstone principle that permitted the development of modern soil mechanics and geotechnical engineering.

Total stresses at a point can be calculated by using stress distribution theory for applied loads, plus the stresses caused by the weight of soil system itself. Static water pressures can be calculated from static water levels. Seepage water pressures can be determined with the aid of flow nets. Therefore, effective stresses can be calculated as the difference of total stress and pore water pressure.

Stress Strain and Compressibility

Since the deformation of the soil depends on the effective stress, deformation of a soil poorly drained may be very strongly time dependent. The applied load on a saturated clay initially is taken up by the pore pressures, and only when drainage can occur will the soil compress. Many clays are saturated and therefore behave in this manner. A partially saturated clay will initially compress the air voids; then air and water drainage will occur.

The amount of compressive strain is a function of the initial effective stress level of the clay ("consolidation pressure"), the magnitude of the applied stress, and stress history and plasticity of the soil.

For normally consolidated soils (those with effective stress levels that have never been greater than present levels), compressibility may be estimated from the Atterberg Liquid Limit of the soil. Otherwise, a laboratory consolidation test (ASTM Designation D2435) is required. A plate load test on a clay soil will not reveal the strain characteristics since the deformation will not be complete until pore pressure dissipate, which may require months, years or even decades.

Granular soils, gravels, and sands, dissipate pore pressure readily and therefore their compression behavior is not so strongly time dependent. At a given confining stress, the stress strain curve of a granular soil is not linear. Nevertheless, it is useful to define and "initial tangent modulus." This modulus increases linearly with confining stress if the soil is loose and is nonlinear for a dense sand. Therefore, for estimates of soil settlements, size and position of the loaded area, the confining stress level, the relative density of the granular soil, and the magitude of the load are generally combined in simiempirical equations. The input is generally (1) relative density of the soil as determined from a standard penetration test, (2) the width of the footing or loaded area, (3) the position of the water table, (4) the depth of surcharge of soil above the bottom of the foundation, and (5) the foundation stress.

Plate load tests may be conducted to determine a stress-settlement curve. At the same stress level, a wide footing will settle considerably more than a narrow plate, up to four times the plate settlement. A 1 ft^2 plate significantly stresses the soil deposit only to the depth of about 1 or 2 ft. The wider the footing stresses the soil proportionately deeper, resulting in greater deformation.

Shear Strength

Soil generally fails in shear, with effective stresses governing the shearing behavior. Therefore, a knowledge of soil drainage conditions is critical to the understanding of its behavior; and granular soils, which readily drain, behave differently from soils with low permeability.

The shearing behavior of granular soils is essentially frictional. Given an initial soil density, shear resistance is proportional to confining stress. This relationship defines the tangent of the friction angle of the soil. The greater the initial density the greater the friction angle, but proportionality between shear resistance and confining stress remains.

Field determination of strength may be accomplished with a static (Dutch) cone test or the friction angle of sands estimated by the results of the standard penetration test (ASTM Designation D1568).

Generally, undrained shearing behavior is critical for cohesive soils. The strength of a cohesive soil is a function of initial effective stresses (consolidation stress), mineralogy of the soil, and stress history. Field shear tests may be performed by the field vane test, and the static cone penetrometer test has found increasing use.

Shrinkage and Swelling

The presence of a highly plastic clay mineral, primarily montmorillonite, gives a clay the property of shrinkage upon drying and swelling upon access to water. The effect is often seasonal, and in many parts of the world affects soils to a depth of up to 20 feet or more. A high Atterberg Plasticity Index is an indicator of swelling potential. Soils compacted on the dry side of the so-called "optimum moisture content" tend to swell, while those compacted at a moisture content above the optimum tend to shrink. The admixture of slaked lime to the soil will retard shrinkage and swelling in many cases.

Compacted Soil

Compaction is the reduction of air voids by mechanical manipulation. It involves both the densification of the soil and a separate control of moisture content. There is a single moisture content at which a soil compacted under a constant compactive effort will achieve a maximum density. Proper compaction control generally involves achieving some percentage of the density of the same soil compacted in the laboratory with a fixed procedure—the so-called Proctor test (ASTM Designation D698). A narrow range of moisture content is also specified. At a moisture content near optimum, increasing density increases strength and decreases both compressibility and permeability. At constant dry density, increasing the moisture content from optimum decreases strength and permeability and increases compressibility. Obviously, compaction control requires very strict inspection to optimize soil behavior.

Uses

Soils in their natural condition or placed artificially can act as a structural component, receiving loads from structures through slabs and footings and distributing these stresses. Soils, generally placed by man, can become drainage layers to carry water flow or the opposite, a layer to prevent substantial water flow.

Soil may be incorporated into housing, as with adobe or as in sun-baked clay construction or a sod house. Such uses tend to be temporary unless methods are devised to preserve strict moisture content control. Soil in such an application may also be vulnerable to burrowing animals and insects.

When designing foundations on soil, it must be first determined whether the soil is granular or cohesive. If granular, it is common practice to design foundations on the basis of the standard penetration test of the Dutch cone, with applicable semiempirical design for both bearing capacity and settlement. Bearing capacity is the ability of the soil to resist rupture (shear failure along some critical arc). Such failure depends on (1) the width of the foundation, (2) the relative density of the soil in the zone one to two times the footing width beneath the foundation, (3) the position of the water table, and (4) the depth of surcharge around the foundation.

Foundations on clay are generally designed on the basis of unconfined compression tests (ASTM Designation D2166) on "undisturbed" samples recovered from soil borings (see ASTM Designation D1587). Field vane tests and cone penetrometer tests are also commonly used. When the stress applied at the base of a footing is less than the unconfined compressive strength of the soil the design is generally safe for bearing capacity, and generally the preconsolidation stress of that soil has not been exceeded and therefore settlements induced by that load will be in the range of 1 in. or less, which is tolerable for most light buildings.

The placement of footing foundations on embankments can create problems. Unless the embankment is well designed and its construction well controlled, the embankment soil may settle under its own weight and that of the building load. Moreover, the weight of the embankment alone often causes large and often uneven settlements of the underlying soil. It is possible that the embankment may overstress the underlying soil causing a bearing-capacity failure.

Soil mechanics and geotechnical engineering is a very specialized field, requiring a specialized graduate education and experience. A competent soils engineer should be available at the very inception of the housing project, especially before the site is finally selected. The necessity for a pile or pier foundation to bypass soil layers not suitable to carry the building loads is very often more costly than the purchase of a more expensive property where the soil conditions allow the use of inexpensive footing foundations.

Choice of Building Materials

Economy is obviously a primary factor to be considered in choosing materials for lower-cost housing. Materials costs may represent as much as

70% of the total construction cost in underdeveloped countries, but as mentioned previously, relative costs of materials fluctuate widely depending on the availability of raw materials, degree of industrial development, local labor costs, and construction methods in common use. Consequently, the choice must be made on a job-to-job basis.

Some general comparisons can be made. Wood is one of the easiest materials to work with, connections between members are easily made and the strength-to-weight ratio is good. It is an attractive material with excellent thermal insulating properties. However, durability and fire resistance are poor in most cases, unless the wood is treated, and stiffness is low.

Concrete is generally durable, fire resistant, and reasonably strong, but heavier than wood. Insulating qualities are good, and by texturing, attractive surfaces can be produced. Required formwork may be complex, however, and greater skill is necessary in construction. Cracking must be controlled through jointing and proper use of reinforcing materials. High-quality Portland cement and a supply of suitable aggregates must be available.

Metals are strong, durable, and possess good dimensional stability. Insulating properties are poor and adequate fire protection must be designed into the structure. A well-developed industrial capability is necessary and connections may require onsite welding, riveting, or bolting.

Plastic building components can be manufactured with moderate to high strength-to-weight ratios and excellent durability. Insulating properties are good and a wide variety of attractive textures and colors is available. Dimensional stability and fire resistance are generally poor, however, and the availability of plastics depends on the availability of the necessary petrochemicals for raw materials and manufacturing facilities.

If industrialized housing construction is feasible in an area, the possibilities for combining different materials in prefabricated units are almost limitless. Sandwich panels using foamed plastic cores and aluminum or steel facings are but one example of the methods used to combine the desirable characteristics of two or more materials.

3

Classical Construction Methods for Low-Rise, Low-Cost Housing

Douglas Wren

The Importance of Studying the Development of Building Methods

Besides their other skills, the Romans were good road builders. Many of their roads exist today, a silent testimony to their soundly based engineering principles. The Romans had two basic rules: (1) provide good drainage and (2) lay down a good foundation of flat stones. (See Figure 3.1.)

Later settlers in England allowed the road system to deteriorate, and some medieval roads were up to a mile wide to allow people to go around the mud. Not until the early nineteenth century did Telford and McAdam get English roads back to these first principles. Part of the reason was political. The coming of the railways soon put an end to good road building again.

The French "chaussée," with "accotements," is similar to a Roman road, and the French reestablished good road building methods about a century before the English.

We periodically forget the first principles of good building and as a result houses often leak, burn down, have roofs blown off, slide down hills, and get flooded with water. Despite the advice of Roman architect Marcus Vitruvius, himself indebted to Greek authors, we often face our houses so that they are hot in summer and cold in winter. Finally, when forced to build large numbers of shelters, either along the ground or piled up in the air, we manage to obtain monumental monotony.

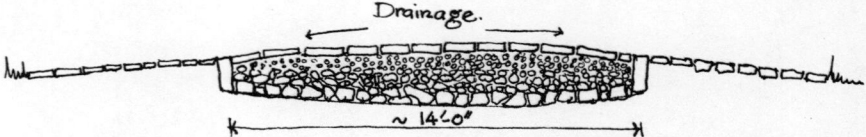

FIGURE 3.1. The construction of a Roman road.

Even with their obvious limitations, classical construction methods can often suggest solutions that are elegant in the extreme, with "elegant" meaning more with less—less material, less effort, and fewer errors. For truly low-cost housing this is a real challenge, because frequently the people have little more than their own hands with which to build.

Energy and Work

The radiation reaching the earth from the sun reacts upon the earth's surface and atmosphere to produce climates and biomes, which in their turn contribute a biomass. We can calculate the total energy available in a given biomass, including all the plants, vertebrates, and invertebrates, it supports.

Man's thumb has enabled him to fashion and use tools, to harness the energy of draft oxen, and to make machines to harness the energy of water (see Figure 3.2). But throughout history, mankind has used, to a very large degree, his own energy and work to produce his shelter.

Many hands are without useful work and many people are without shelter. It would seem logical to try to develop systems that will enable some people to build for themselves.

Over millions of years, solar radiation, through photosynthesis in plants, has been stored as nonrenewable energy in the form of fossil fuels. The energy requirements of many modern materials are very high—even for the

FIGURE 3.2. Energy and work.

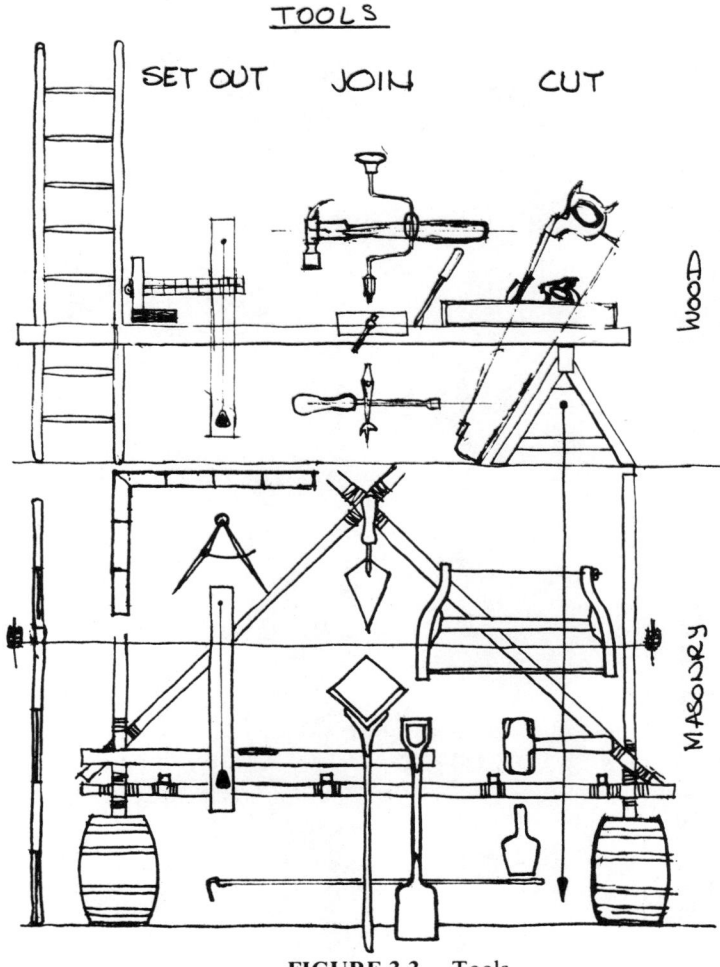

FIGURE 3.3. Tools.

manufacture of such staples as steel, cement, and crushed stone. (See Figures 3.3 to 3.5.)

Roof Construction

Flat and Low Slope Roofs

Timber sizes limit spans to about 16 ft for flat and low slope roof construction (see Figure 3.6). Bearing-wall structural systems respond very well to roof systems that apply well-distributed loads vertically.

SOURCES OF MATERIALS WITH EXAMPLES OF THEIR USE

FUNCTION	INDIGENOUS					INDUSTRIAL		
Load bearing walls	Clay/sand Soils. (adobe)	Stone (wall units)	Plant fibres (straw bales)	Wood (studs)	Burnt clay products (bricks)	Concrete products (blocks)	Metal Products	
Cimentitious & joining materials	Clay soil (mortar)		Plant fibres (rope)		Burnt limestone (lime)	Burnt limestone (Portland cement)		Bitumens Elastomers & plastics. (adhesives)(Caulkings & glues.)
Horizontal load bearing elements.				Wood (Joists & beams)	Plywood (flooring) Glulam (beams)	Concrete & steel (Floor slabs)	Steel (Beams, Joists, decks)	
Roof Coverings, sloped.	Clay Soil (Prairie Sod roofs)	Slate, Stone. (Roofing units)	Wood (Shingles) Plant fibres (thatch)		Burnt clay (Roofing tiles)		Steel (Sheet metal roofing)	Bitumens & wood fibre. (Shingles) Sheet plastics. Fiberglass (Panels)
Roof Coverings, flat								Bitumens (Built up roofing) Plastics (P.V.C. roofing)
Windows					Limestone, Sand, Soda ash (glass)			Acrylic Plastic domes

FIGURE 3.4. Building materials and their uses.

Pitched Roofs with Substantial Slopes 39

In Figure 3.7 the elevation the roof joist is represented by three parallel lines to show that it is acting in bending. The section shows that loads are well distributed along the wall. No lateral forces tend to overturn the wall.

When the span exceeds 16 ft or thereabouts, the designer must use a series of beams, which may be supported on bearing walls or posts. Figure 3.8 illustrates the no-thrust principle with an inclined roof. The beams are shown in black. The roof consists of solid planking instead of joists and plywood.

As Figure 3.9 shows, a low slope roof can also be constructed using load-bearing ceiling joists as support. Further, the structural analysis depicted in Figure 3.10 shows that both the joists and rafters are beams acting in bending. The compression strut in the middle is shown by a double line.

A conveniently positioned bearing wall can also be used to support roof members. Figure 3.11 illustrates how.

The shed roofs in Figure 3.12 are being used for a tile roof covering. The project shows staggered terrace housing of $1\frac{1}{2}$ stories. The inclined rafters must be securely supported at both ends, otherwise they will exercise a horizontal thrust, somewhat like a ladder placed against a wall. And the ladder, as shown in Figure 3.13, will fall if the coefficient of friction between the foot of the ladder and the soil is too low, or if the thrust is too great, because of a shallow angle.

Pitched Roofs with Substantial Slopes

NO-THRUST SYSTEMS In the no-thrust pitched roof in Figure 3.14 are closely spaced inclined beams called "rafters." Both beams and rafters should be joined together over supports.

Figure 3.15 illustrates how roofs should always be firmly attached to the walls. Where winds are very strong, metal anchors must be used to avoid uplift.

Structures such as the one in Figure 3.16 are only an extension of the principles of Stonehenge.

The usable space in Figure 3.17 is created in the roof volume. Again, the system does not tend to push the walls over. There is no horizontal thrust. The framing for windows in the roof is costly.

SYSTEMS WITH HORIZONTAL THRUST The great disadvantage of the flat or inclined beam system is that it requires large timbers for all but the smallest spaces. Since in structures one tries to cover the most space with the least material, one must soon turn to a system of two inclined beams or rafters supported against each other. To prevent the walls being pushed over, the ends of the rafters are tied together with the ceiling joists.

House Constructed in Canada in 1975			House Constructed in Great Britain in 1912		
	Weight of Materials Produced within 30 Miles			Weight of Materials Produced within 30 Miles	
Materials	Tons	%	Materials	Tons	%
960 ft², including walls			910 ft², including walls		
1 story, with basement			2 stories, without basement		
Wood walls and exterior finish			Brick walls and exterior finish		
Wood floor and roof framing			Wood floor and roof framing		
Asphalt shingle roof			Clay tile roof		
Plasterboard interior finish			Lime plaster interior finish on wood lath		
Climate—30 to 85°F			Climate 20 to 80°F		
The Canadian house has a complete warm air heating system, electricity, full bathroom, hot and cold water			The English house had a cast iron cooking stove, copper wash boiler, sink, cold water. One fireplace downstairs, one upstairs. Electricity is not mentioned. Oil lamps were probably used		

Crushed stone and sand				
(70 tons of crushed stone, 28 sand)	98	98	Ballast sand and gravel	87 87
Portland cement	0	0	Brick	95 95
Portland cement	12	12	Portland cement	14 14
Gypsum (incorporated into gypsumboard)	0	0	Lime	7 7
Wood and wood products	6	0		0 0
	$9\frac{1}{2}$	7	Wood and wood products	9 0
Metals	2	0	Metals	1 0
Asphalt products	$1\frac{1}{4}$	0		0 0
Plastic products	$\frac{1}{4}$	0		0 0
Glass	$\frac{3}{8}$	0	Glass	$\frac{1}{4}$ $\frac{1}{4}$
Insulating materials	$\frac{3}{4}$	0		0 0
	0	0	Roofing tiles	8 8
Total weight of materials	130 tons	90%		221 tons 95%

FIG. 3.5. The importance of using local materials.

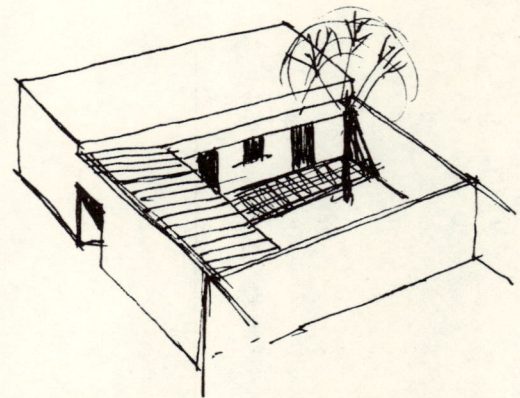

FIGURE 3.6. A Middle East courtyard type house.

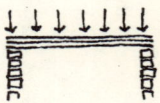 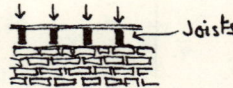

FIGURE 3.7. The distribution of loads on a bearing wall from a flat roof constructed with joists.

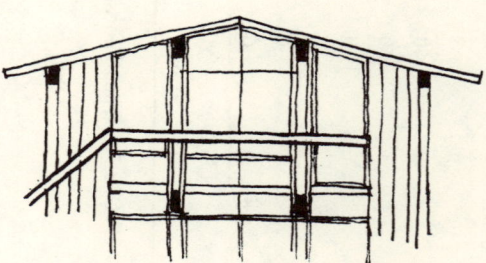

FIGURE 3.8. Post and beam construction.

 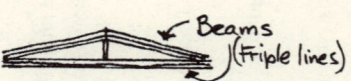

FIGURE 3.9. Structural ceiling joists as a roof support system.

FIGURE 3.10. A structural analysis of the roof shown in Figure 3.9.

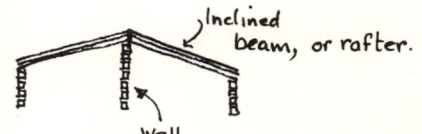

FIGURE 3.11. The no-thrust principle, using a central bearing wall as support.

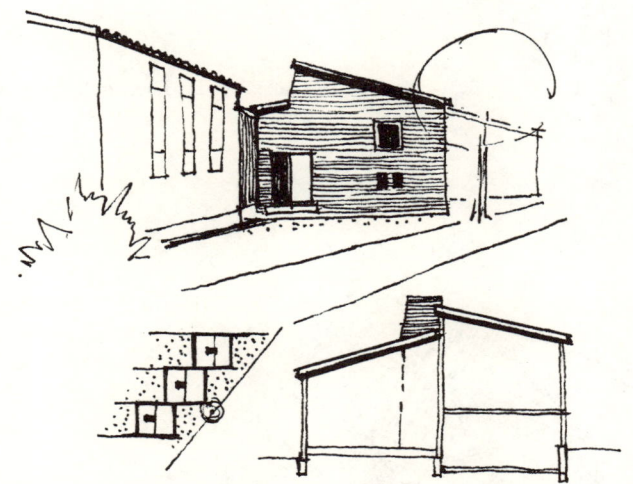

FIGURE 3.12. Staggered terrace housing with no-thrust roofs.

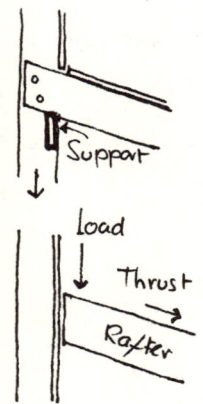

FIGURE 3.13. The inclined rafters of shed roofs must be supported at both ends.

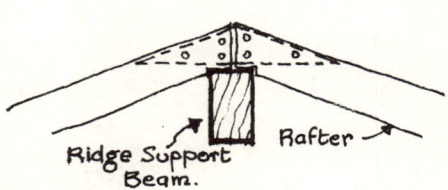

FIGURE 3.14. A no-thrust pitched roof.

43

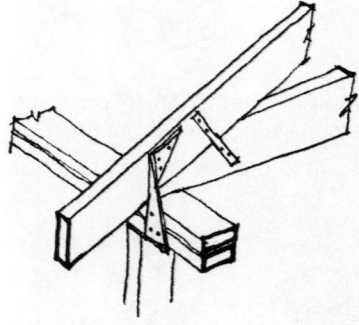

FIGURE 3.15. The anchorage of the roof to the walls.

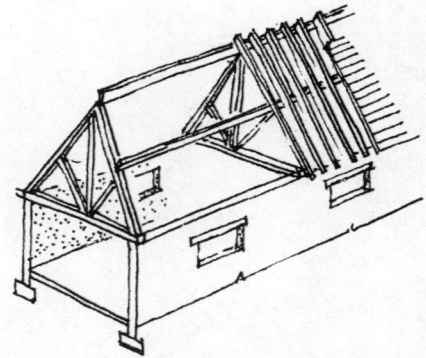

FIGURE 3.16. The use of purlins avoids bending moments in the principle rafters.

The framing for windows is costly in roof spaces.

FIGURE 3.17. The forerunner of the Gambrel or Mansard roof.

Pitched Roofs with Substantial Slopes 45

FIGURE 3.18. The traditional construction of buildings in much of the Third World.

A simple concrete block dwelling with galvanized iron or corrugated asbestos cement roofing.

For 50 lb/ft² snow loads, rafter span depth ratios are 1:24 rising to 1:36 for loads of less than 10 lb/ft², such as galvanized iron sheeting in sheltered areas. Ceiling joists have span-to-depth ratios of 1:32 if plastered and 1:36 if covered with a more flexible material. Buildings up to 20 ft wide can therefore be constructed, depending on the climatic and roof covering requirements, using 2 × 4 in. timbers. (See figures 3.19 and 3.20.)

If a reasonably centralized bearing wall exists as in Figure 3.21, the constructor may strut off the wall to support the rafters of the roof frame and thus reduce the span and size of the rafters. The tiles, as shown in Figure 3.22, may act in tension or compression, depending on the roof loads.

To create more space, ceiling joists were sometimes raised to form collar ties as shown in the eighteenth-century New England "salt box" with lean-to addition in Figure 3.23. Keeping the height down made building easier, but at the risk of a less rigid roof and wall assembly.

Gable ends using hip rafters are very stable. They avoid the scaffolding problems associated with masonry gable end walls (see Figure 3.24).

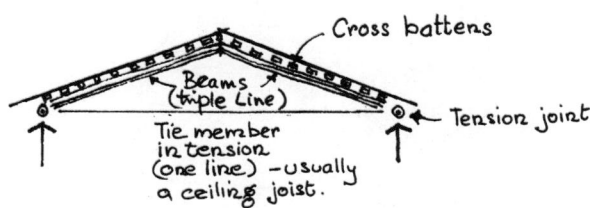

FIGURE 3.19. The triangle is a stable structural form, and engineers frequently use triangulation in their designs.

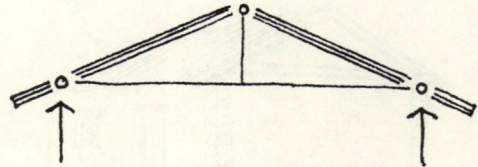

FIGURE 3.20. Metal tie rods and sag rods are used to resist the lateral thrust of the rafters if ceiling joists are not required.

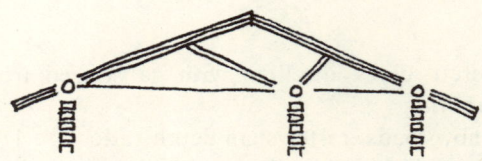

FIGURE 3.21. Strutting off a central bearing wall to reduce the rafter span and therefore its dimension.

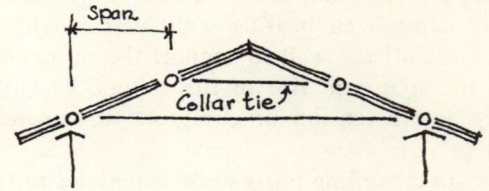

FIGURE 3.22. Collar ties reduce the rafter span.

FIGURE 3.23. The eighteenth-century New England "salt box."

Pitched Roofs with Substantial Slopes   47

FIGURE 3.24.   A hipped roof.

Of the two types of roofs shown in Figure 3.25, the conical roofs are ideal for weak load-bearing walls, because loads are well distributed. Pyramidal roofs using hip rafters require complex tension joints. All insulated ceilings for cooler climates are hard to construct. For larger floor areas, both forms require a disproportionate increase in material.

Roof trusses, using members in compression and tension only, are economical when the roofing material has structural properties. See Figure 3.26.

An A-frame construction, as in Figure 3.27, creates living space totally within the roof. The horizontal thrust is low.

With a rise of 9 on 12 and a central load of 6 units, the ceiling tension member will carry a load of 4 units. See Figure 3.28.

When the pitch is reduced to 3 on 12, the tension load in the ceiling joist rises to 12 units, and a very large number of nails would be required to secure the joint—too many in fact. See Figure 3.29.

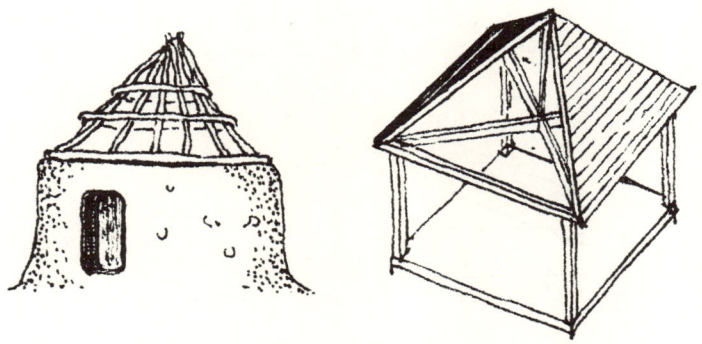

FIGURE 3.25.   Conical and pyramidal roofs.

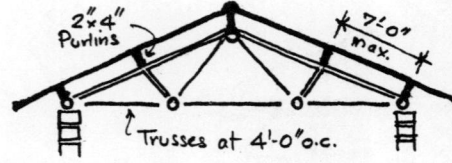

FIGURE 3.26. A roof truss to carry light loads. Can be spaced at 4 ft oc with asbestos cement corrugated sheet roofing.

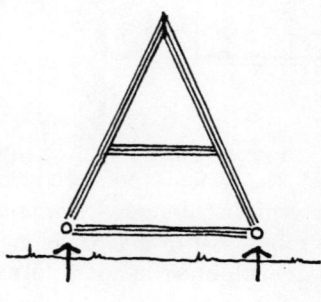

FIGURE 3.27. An A-frame construction.

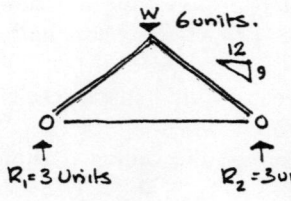

FIGURE 3.28. Horizontal forces in a ceiling joist of triangulated roof construction with a substantial slope.

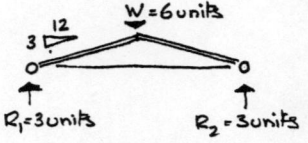

FIGURE 3.29. Increased horizontal forces in a ceiling joist of a triangulated, low-slope roof.

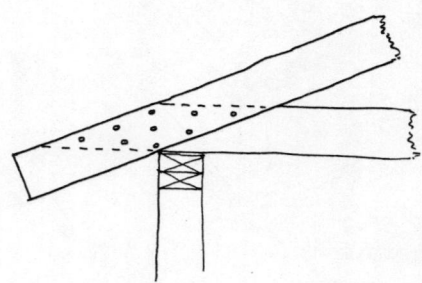

FIGURE 3.30. The anchorage of the ceiling joist to the rafter with nails.

Pitched Roofs with Substantial Slopes   49

Roofs are therefore restricted to 4:12 or greater for triangulated truss and frame systems. See Figure 3.30.

Figure 3.31 shows how the tusk of the tusk-lemon joint must be placed in the middle third or "no-tension" zone. Tension joints, in fact, are difficult to design in wood. The traditional mortice and tenon joint with hardwood pins removed as much as two-thirds of the wood from the end of the member. This meant that roof members were much larger than was strictly necessary for the spans.

With the arrival of the cheap wire nail on the market in the nineteenth century shown in Figure 3.32, the construction industry was profoundly changed. Nailed connections became economical, and the smaller untapered nails split the wood less often.

Large trusses are spaced widely apart. Trusses spaced at 2 ft oc are called "trussed rafters." With 2 × 4 in. wooden members and ½ in. plywood gussets, the roof loading would be 30 lb/ft². See Figure 3.33.

In Figure 3.34 the rafters must resist bending moments and are shown with triple lines. Struts or compression members are shown with double lines, and tension members with single lines. Tension joints are indicated with circles. Trussed rafters are spaced at 2 ft oc. Spans ranging from 16 to 28 ft are common.

First conceived with plywood nailed gussets, most light-trussed rafters are now made with a form of gang-nail metal plate that enables rapid prefabrication of the trusses. For larger spans, split ring (shear) connectors with bolts are used. See Figure 3.35.

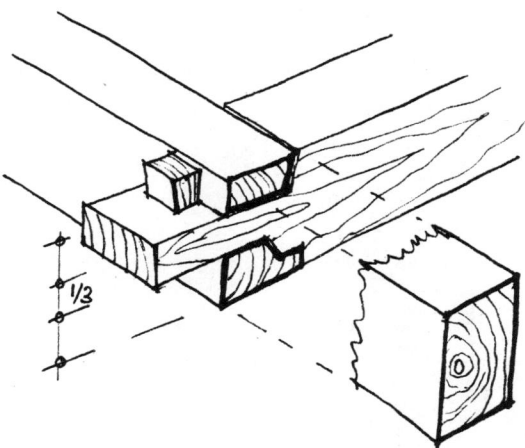

FIGURE 3.31.   A tusk-tenon joint.

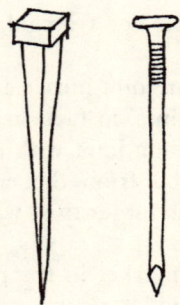

FIGURE 3.32. The evolution of the nail.

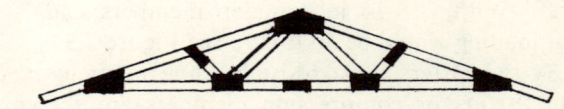

FIGURE 3.33. Inclined trussed rafters.

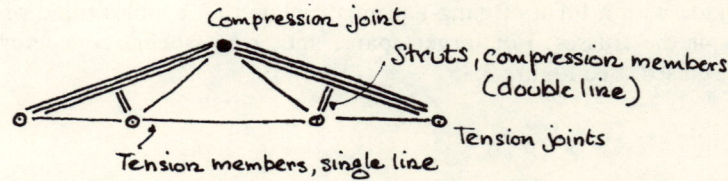

FIGURE 3.34. A structural analysis of the inclined trussed rafter.

FIGURE 3.35. Connectors—a gang-nail metal plate and a split ring.

Wood Framing for Flat Roofs and Floors

Joists, beams, or horizontal poles are the primary structural members (see Figure 3.36). In wood, maximum spans are limited to about 16 ft, the economic maximum length of softwoods. (Ratios quoted are for 40 lb/ft^2 loading and $1/360$ deflection).

Planks, or sheets of plywood are the secondary members. The strength of these members determines the joist spacing. Secondary members have a depth-to-span ratio of 1:32. Wooden beams supporting a roof have a ratio of about 1:16. See figures 3.37 and 3.38.

In construction, the floor framing must obviously resist vertical loads, less obviously the floor resists twisting in the horizontal plane (see Figure 3.39).

Load-sharing wood joist systems have a depth-to-span ratio of 1:20 (Figure 3.40). And steel beams supporting a roof or a floor have a depth-to-span ratio of 1:20. Wooden beams supporting a roof and one floor have a ratio of 1:12 (Figure 3.41).

By using joist hangers, the clearance under a beam can be increased (Figure 3.42).

Joists are doubled up around openings (figures 3.43 to 3.45).

Floor joist must be in place before backfilling (Figure 3.46).

Vertical Loads

Bearing-Wall Construction

Bearing walls are usually divided into three categories:

1. Gravity-bearing walls.
2. Steel-reinforced bearing walls.
3. Closely spaced pole or stud walls.

GRAVITY-BEARING WALLS These walls are constructed with masonry units, plain concrete, rammed earth, sod, straw bales, or similar materials. They cannot accept tension anywhere in the system and cannot act in bending. (See Figure 3.47.) All loads must be confined to the middle third of the wall.

STEEL REINFORCED BEARING WALLS Steel reinforcing is introduced into bearing walls to permit the wall to act in bending or to resist buckling.

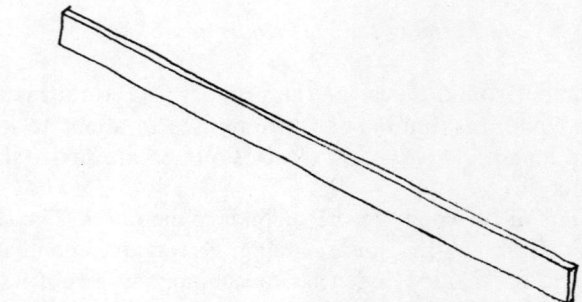

FIGURE 3.36. Primary horizontal members.

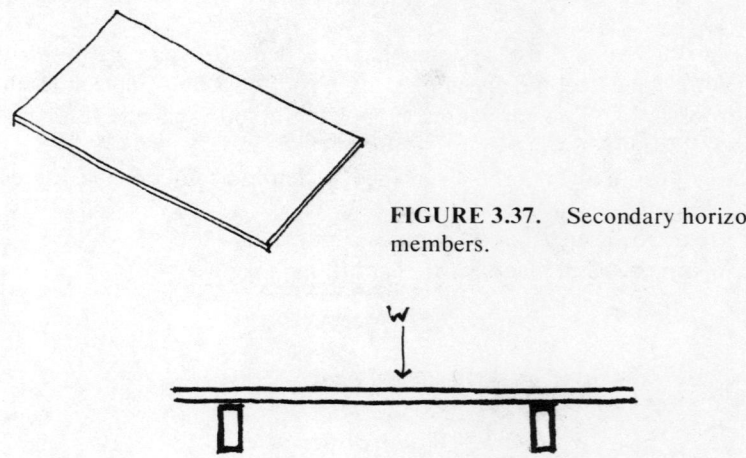

FIGURE 3.37. Secondary horizontal members.

FIGURE 3.38. Determining joist spacing.

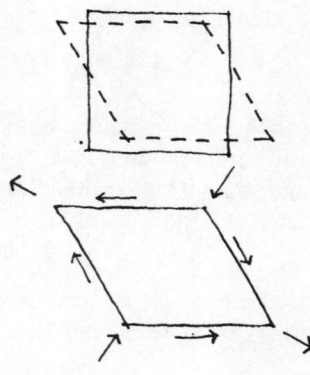

FIGURE 3.39. The floor as an anti-torsion diaphragm.

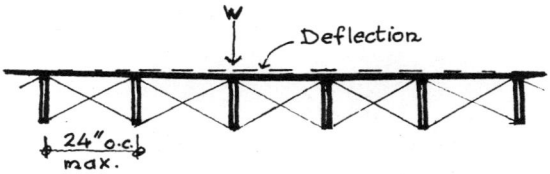

FIGURE 3.40. Section through a wood joist floor.

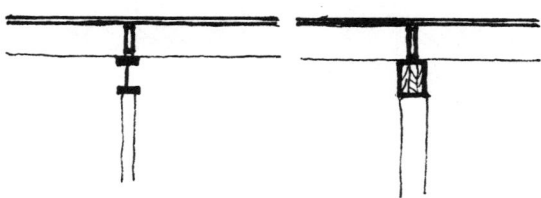

FIGURE 3.41. The depth-to-span ratio of beams.

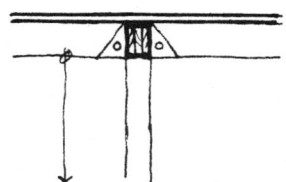

FIGURE 3.42. Maintaining useful head clearance under a beam.

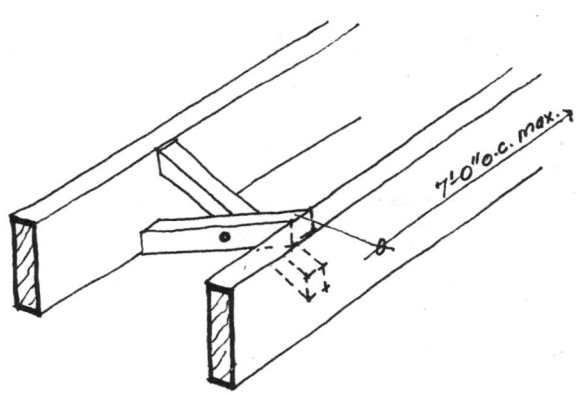

FIGURE 3.43. Blocking or diagonal bracing stops twisting and increases load-carrying capacity.

53

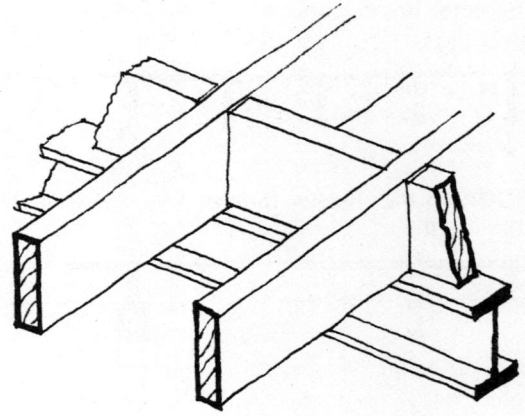

FIGURE 3.44. Solid blocking over supports prevents joists from twisting.

FIGURE 3.45. Floor framing around openings.

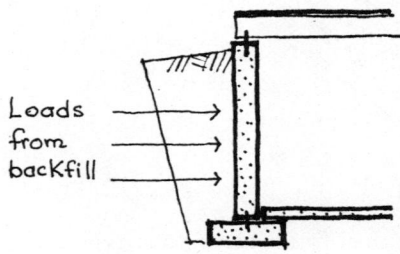

FIGURE 3.46. Floors are used to resist horizontal thrust on basement walls.

Walls built of concrete, brick, concrete block, and even adobe, can all be reinforced with steel bars.

CLOSELY SPACED POLE OR STUD WALLS   Walls are produced by placing poles and studs close to each other.

BUCKLING   Walls of closely spaced vertical wood poles or studs spaced at not more than 24 in. oc. usually fail by buckling rather than crushing, and the lengths of the members are largely determined by the size of the member. (See Figure 3.48.)

CRUSHING   While for both categories the crushing of the materials because of loads can occur, it is usually a problem confined to gravity-bearing walls, due in part to the self-weight of the wall. As can be seen from the table, walls of rectangular units are stronger than walls of, say, rounded pebbles or rounded cob balls. It is self-evident that walls of squared stone, brick, or concrete block, will be more resistant to crushing than earth walls.

LOAD DISTRIBUTION   All load-bearing systems perform better when loads are evenly distributed rather than concentrated. For this reason, it is preferable to avoid large spans with widely spaced girders or large openings that necessitate lintels and inevitably concentrate loads. See Figure 3.49.

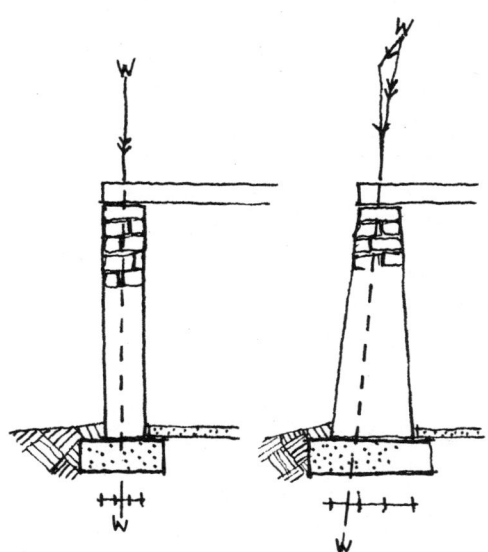

FIGURE 3.47.   Forces in a masonry wall.

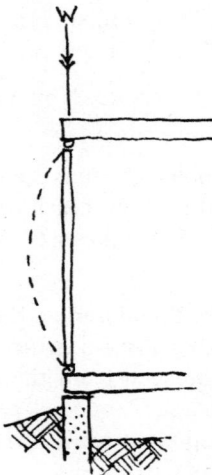

FIGURE 3.48. Buckling in a wood stud wall.

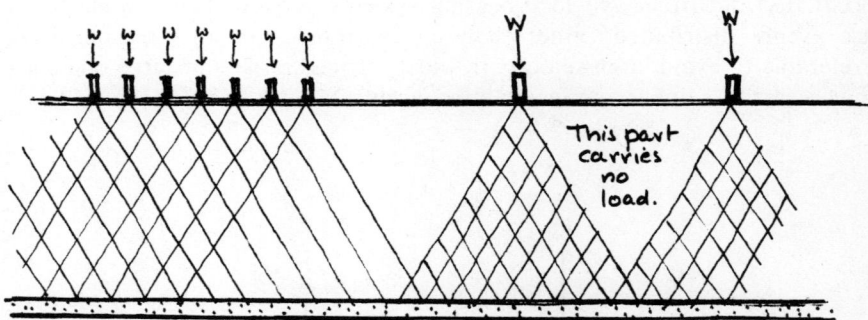

FIGURE 3.49. A comparison of the load distribution from closely and widely spaced structural members supported by a masonry-bearing wall.

Reinforced concrete bond beams are commonly used as a means of distributing loads in masonry walls. In countries where Portland cement and steel are in short supply, a similar bonding and load-distribution function can be performed by one or two courses of large through-the-wall units under floor and roof members. See Figure 3.50.

Provided that there is adequate lateral abutment, lintels over small openings in masonry walls support only the 60° isosceles triangle of masonry shown as a white space in Figure 3.51. This is because of arch action. It is important to avoid concentrated floor loads in this triangular space, which means that in practical applications, the principle can only be exploited when the floor joints are parallel to the wall surface.

Vertical Loads 57

LATERAL SUPPORT   In all cases cited, it is assumed that the wall will be supported at each floor level by the floor construction itself and at roof level by a roof that does not exert a lateral force on the wall. Parapet walls and garden walls do not benefit from lateral support at the top, and their height-thickness ratio is extremely low (see Figure 3.52).

THICKNESS   No method exists for determining the thickness of load-bearing walls according to the loads to be carried and the materials used. Nevertheless, Figure 3.53 gives some idea of commonly encountered proportions. As the height of a masonry wall increases, so does the load per square foot on the masonry. If the designer wishes to maintain a uniform loading, he or she must gradually increase the wall thickness toward the base or opt for a partial solution of increasing the wall thickness at successive floor levels. In wood frame construction, another option exists: spacing the studs more

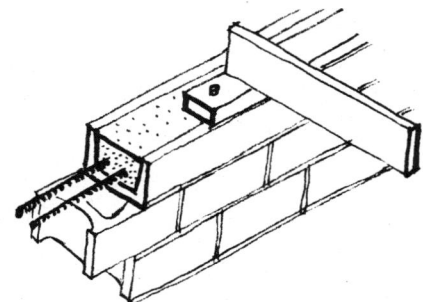

FIGURE 3.50.   A reinforced concrete bond beam.

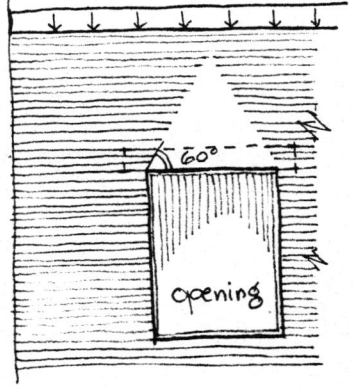

FIGURE 3.51.   Arch action over openings in masonry walls.

closely together. (In reinforced concrete walls, designers may maintain uniform wall thicknesses by increasing the percentage of steel).

Post and Beam Construction

Like all columns or posts, the tent pole's bearing capacity depends to a great extent on the relationship between its length and its least dimension. See Figure 3.54.

The effective length, and therefore the load-bearing capacity, can be spectacularly improved by the use of intermediate and diagonal braces. Unfortunately, as can be seen in Figure 3.55 this is more easily done in the plane of the wall. The poles are unbraced on the other axis.

Pole structures in which the poles are embedded in the ground are very efficient structurally (see Figure 3.56). The effective length is reduced, which means the same pole can carry a greater load or a smaller pole can be used. Elevated pole structures are used in areas subjected to seasonal flooding. Curiously, we now try to build on the ground in similar situations and the houses float away in times of severe flooding. If chemically treated the poles should last almost indefinitely.

However, squared lumber is much easier to build with. When only the smaller sizes of wood are available, columns can be built-up or spaced. See figures 3.57 and 3.58.

Figure 3.59 gives an accounting of the ratio of height to least dimension for spruce wood columns.

Horizontal Loads

Dynamic Loads

Both wind and earthquake loads on buildings are complex dynamic forces that depend on many factors. To simplify the calculation for designers, these loads have been translated into simplified static formulas that are quite sufficient for small buildings.

Wind Loads

Wind has become a major problem as buildings have become higher. The exposure of small buildings is a prime consideration. It is fairly obvious that those buildings situated on sea coasts or on open prairie will have to resist much greater loads than similar buildings situated in urban areas. See Figure 3.60.

What is less obvious is that the shape and orientation of the building and particularly the roof angle will materially affect wind loads on the building structure and the building envelope. Negative suction loads often greatly exceed the positive loads, and in fact high winds and tornados lift off roofs by suction. See figures 3.61 to 3.63.

Figure 3.63 shows a wall plate anchored by a long bolt that will involve the dead weight of a large volume of stone masonry and thus provide a positive downward force. For frame walls the sheathing will adequately anchor the frame to the foundation plate, which itself is attached by bolts to the concrete foundation wall. These bolts also prevent the horizontal sliding of the wooden frame on the concrete wall.

For small buildings the horizontal shear due to wind is so small that in the case of 2-story framed houses it can adequately be resisted by the interior gypsum wallboard lining. Masonry walls properly positioned in accordance with the principles of statics will normally have little difficulty in resisting the horizontal shear. Prefabricated wall panels are subject to negative exterior forces on the down wind side to which must be added the interior pressure due to stack effect and temperature differences. Generally speaking, traditional light frame construction shared these loads between an interior and exterior membrane. See Figure 3.64.

A thatched roof is an envelope member, and Irish cottages thus clothed were furnished with ropes and rocks to resist the suction (see Figure 3.65). Chicken wire was also used under the ropes, but this was largely to discourage birds from stealing and resting therein.

Earthquakes

Earthquakes permit stresses to be relieved at the earth's surface. This process will generate forces that the buildings have to resist.

Earthquake loads present a different type problem from that resulting from wind loads since adjoining buildings cannot afford protection as is the case for wind loads. The crust of the earth consists of about 12 giant plates floating on the molten basalt interior (see Figure 3.66). Where these plates collide, tremendous forces are set up. The plate with India on it, has moved under the Asiatic plate, giving rise to the Himalayas. The African plate is drifting northward into the European plate and recently caused severe earthquakes in Northern Italy. The North American plate has major faults on the West Coast and up the St. Lawrence Valley. Recently, the United States launched a 2 ft diameter satellite in the form of a solid brass sphere. Covered with dimpled aluminum, the satellite will serve as a fixed station in

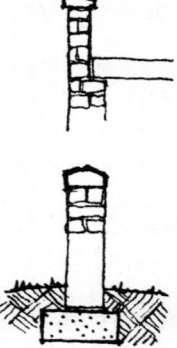

FIGURE 3.52. Parapet walls and garden walls are not supported laterally.

space that will permit accurate laser measurements of the tectonic plate movements.

Earth tremors cause rapid lateral movements of the supporting crust and will topple structures not designed to resist the resultant horizontal shears. The weight of the structure and particularly the roof will materially affect the magnitude of the horizontal shear to be resisted at the base. See figures 3.67 and 3.68.

The traditional Japanese house shown in Figure 3.69 is often cited as being designed to resist earthquakes. But the heavy tile roof commonly used refutes this statement.

Lateral Stability

Requirements for Stability

To be laterally stable a building must resist horizontal shear forces. If these forces are not adequately resisted, racking or collapse will result.

Cantilevers—Columns without exterior guys

A wooden post in the ground resists the wind by acting as a cantilever in bending and by the shear resistance of the wood at the base. If a large sign is placed on top of the post the wind may blow the post over without breaking it. See Figure 3.70.

In Figure 3.71 a ground level beam provides resistance to lateral movement. If the structure were of reinforced concrete, a horizontal slab would resist the overturning movements.

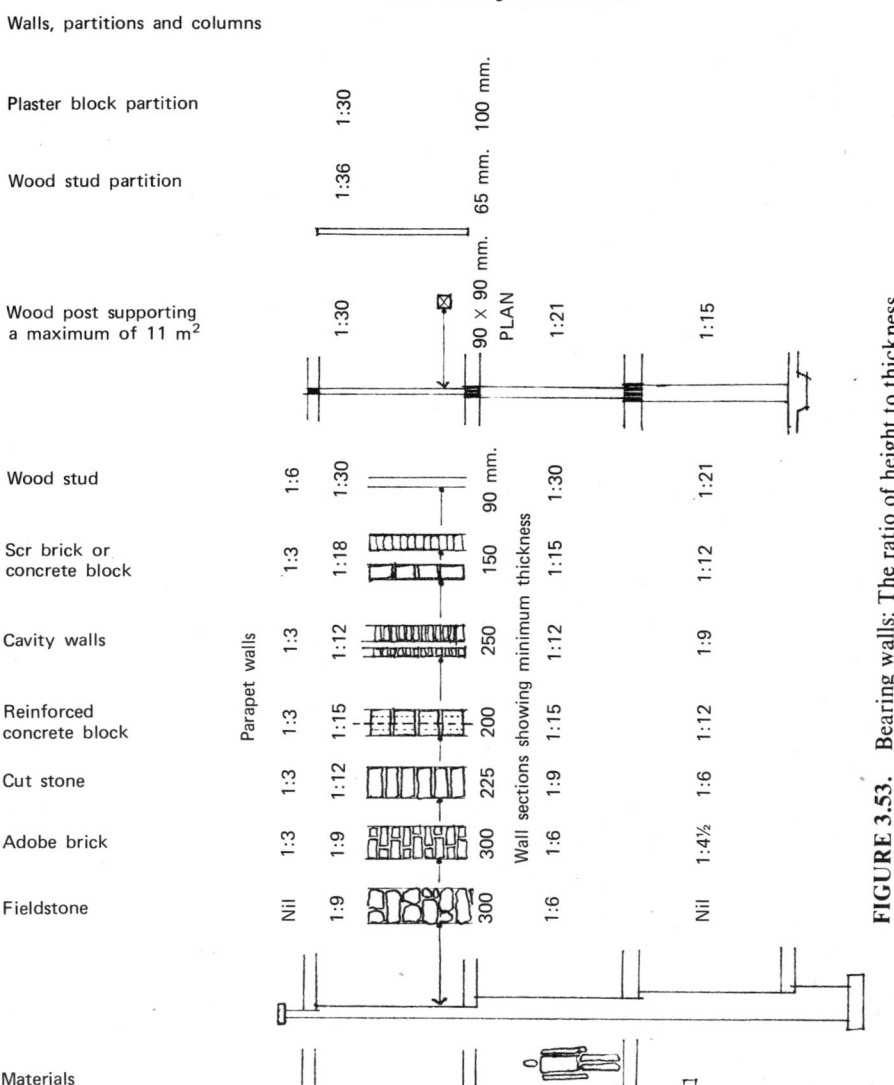

FIGURE 3.53. Bearing walls: The ratio of height to thickness.

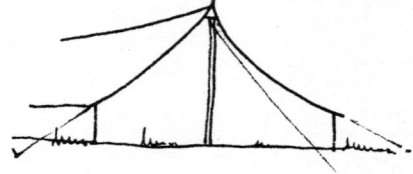

FIGURE 3.54. The tent pole is a simple column.

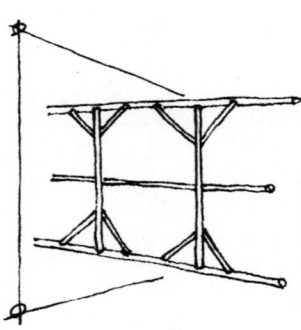

FIGURE 3.55. The effective length of a column.

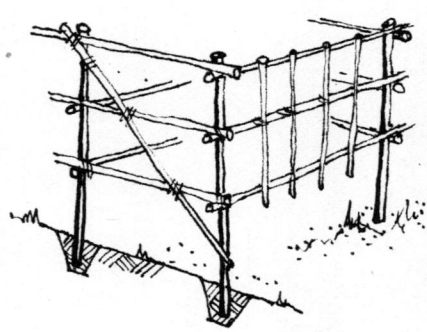

FIGURE 3.56. Lashed-pole construction.

FIGURE 3.57. The squared column, which can be cut from a tree, carries 75% of the load the round tree trunk can carry.

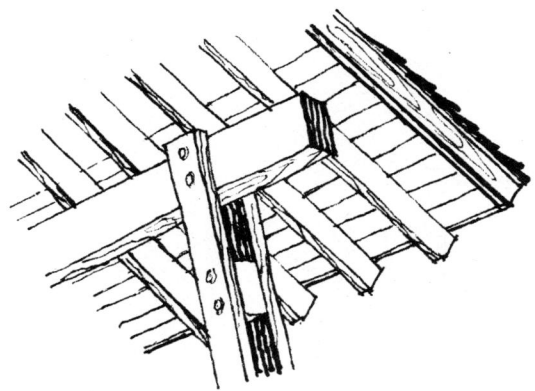

FIGURE 3.58. A typical detail of the post and beam construction commonly used on the West Coast of the United States.

Ratio	Height (ft)	Size (in.)	Load Supported (lb)	
30:1	6	2½ × 2½	Mullions for roof loads	5,000
24:1	8	3½ × 3½	Roof	10,000
18:1	8	5½ × 5½	Roof and one floor	20,000
15:1	9 ft. 6 in.	7½ × 7½	Roof and two floors	30,000

FIG. 3.59. The ratio of height to least dimension for spruce wood columns.

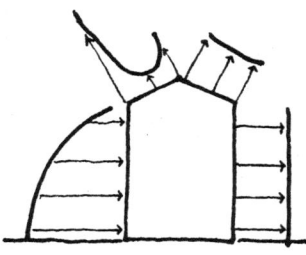

FIGURE 3.60. Positive and negative wind forces (the negative forces are bigger).

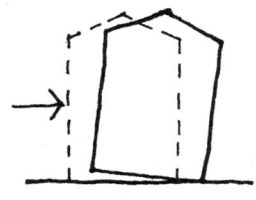

FIGURE 3.61. Unanchored, a building could slide or lift off its foundations.

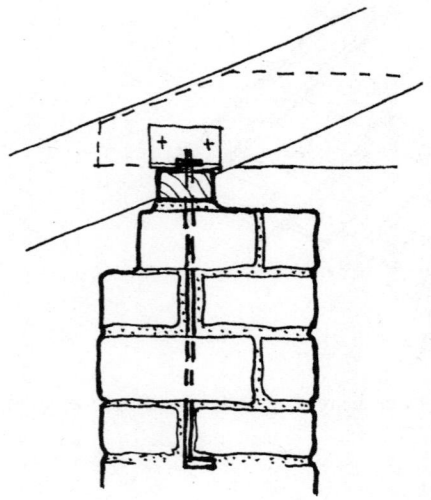

FIGURE 3.62. Light modern roofs of wood and steel must be firmly anchored to the supporting structure.

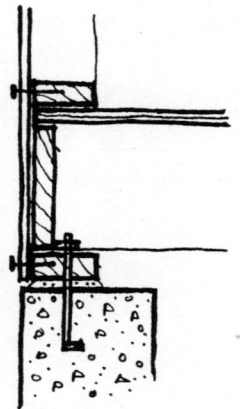

FIGURE 3.63. Anchorage of the wall system to the heavy foundation.

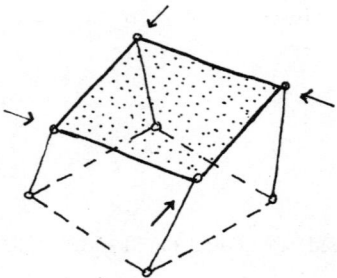

FIGURE 3.64. Wind loads rack and twist a building.

FIGURE 3.65. The thatched roof of a traditional Irish cottage.

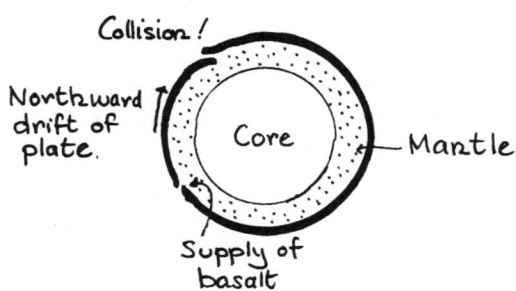

FIGURE 3.66. The crust of the earth consists of a series of solid tectonic plates floating on a molten basalt interior.

FIGURE 3.67. Horizontal earthquake forces acting on a building sitting on a tectonic plate that has collided with its neighbor.

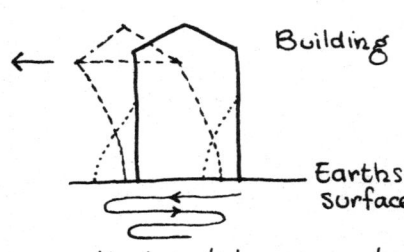

FIGURE 3.68. An enlarged view of what happens to a building when an earthquake occurs.

Heavy Masonry Structures

To resist wind forces only, the coefficient of friction in the mortar joints must resist horizontal sliding. When earthquake forces must be resisted, the walls have to be reinforced with steel to enable them to resist bending movements, and all joints must connect the walls rigidly to the floors. It is common knowledge that classical construction methods have failed to provide adequate resistance to earthquake shocks. This is because seismology is a new science.

Masonry walls must have reinforced concrete cores and bond beams to tie the roof to the wall. The upper part of the section in Figure 3.72 shows reinforced concrete fill between brick wythes, while below, an alternative construction of concrete blocks with filled cores is shown. Foundation depth depends on climatic conditions and soil type. Weak compressible soils with low seismic shear have a factor of 1.5 instead of a factor of 1.0 for normal soils.

Most habitations need some form of heat, at least for cooking meals. A solid masonry fireplace with a tightly filled floor system around it can provide excellent lateral stability (see Figure 3.73). Because of the absence of terra cotta chimney liners, houses often burned down when using the fireplace and chimney for support because the joists were wedged tightly against the masonry. However, Figure 3.74 provides an example where the floor will transmit to the walls the stability provided by the fireplace.

Shear Walls

POSITION OF SHEAR WALLS   If the fireplace is not centrally placed, at least one shear wall must be added. This is necessary for the structural safety of the building. See figures 3.75 to 3.77.

With four walls in solid masonry, this stone house satisfies all the foregoing requirements for lateral stability against wind loads. The coefficient of friction of the mortar joints is the key factor.

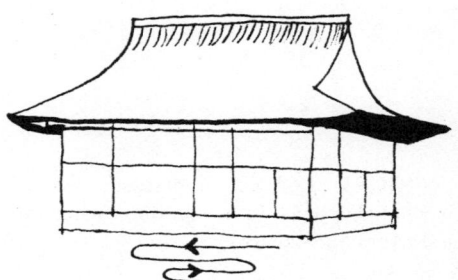

FIGURE 3.69. A traditional Japanese house.

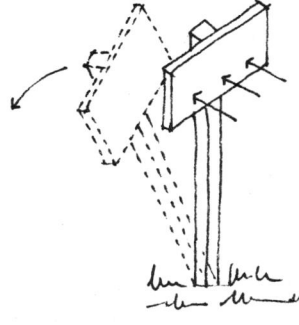

FIGURE 3.70. Wind overturning a wooden sign post.

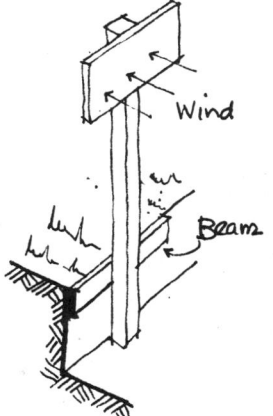

FIGURE 3.71. The lateral stability afforded by a ground beam.

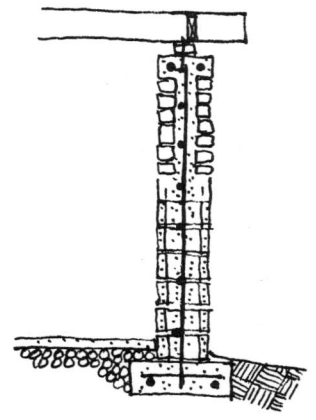

FIGURE 3.72. A reinforced masonry wall.

FIGURE 3.73. A masonry fireplace can provide good lateral support.

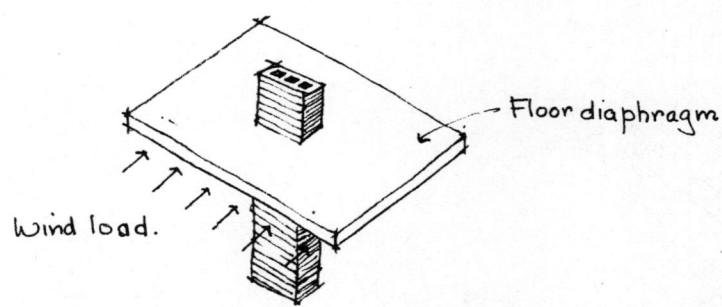

FIGURE 3.74. Wind acting on a building.

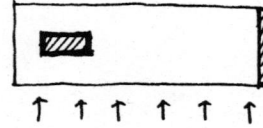

FIGURE 3.75. The wall and massive fireplace must resist horizontal shear.

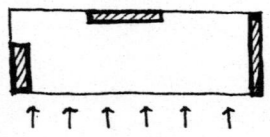

FIGURE 3.76. In more extreme cases, two shear walls would be necessary to provide stability to the structure, plus the fireplace.

Lateral Stability

FIGURE 3.77. A traditional masonry house.

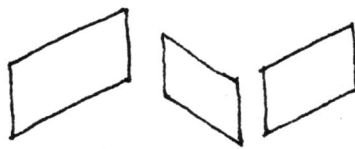

FIGURE 3.78. Rule 1: There must be at least three planes.

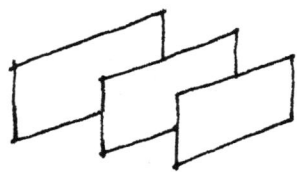

FIGURE 3.79. Rule 2: The three planes must not be parallel to each other.

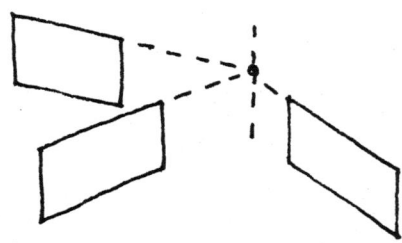

FIGURE 3.80. Rule 3: The three planes must not meet at a single point.

The rules regarding the number and positions of shear walls acting alone are that (1) there must be at least three planes; (2) The planes must be unparalleled, and (3) the planes must not meet at a single point. See figures 3.78 to 3.80.

Post and Beam Structures

The post and beam structural system is inherently unstable and must rely on one of the following methods to insure lateral stability:

1. Triangulation.
2. Heavy masonry core structure.
3. Rigid joists.
4. External guys.
5. Shear walls.

The non-load bearing walls filled with masonry will provide structural effects similar to shear walls. (Figure 3.86)

Suspension Structures

Lateral stability is provided by cable guying systems anchored to the ground.

Vaults and Domes

Vaults may be considered as a series of arches side by side. Domes are a series of concentric arches.

PROPORTIONS OF SHEAR WALLS

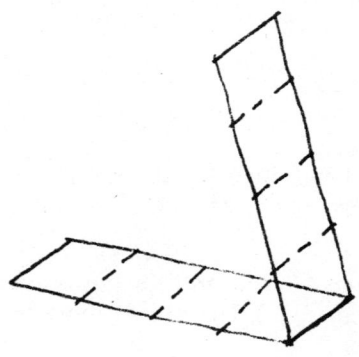

FIGURE 3.81. Wall and floor diaphragms should have ratios not exceeding 1:4.

ANCHORAGE OF LIGHT SHEAR WALLS OR SHEAR DIAPHRAGMS

FIGURE 3.82. Walls and floors must be well anchored to foundation walls or supporting structures.

RIGIDITY OF WOODEN SHEAR WALLS (DIAPHRAGMS)

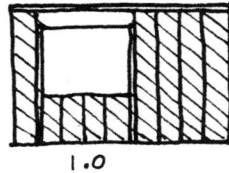

FIGURE 3.83. Rigidity factor 1.0: The diagonal sheathing will rack and twist.

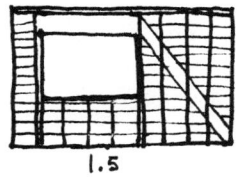

FIGURE 3.84. Rigidity factor 1.5: The cut in diagonal braces on horizontal sheathing stiffens the construction.

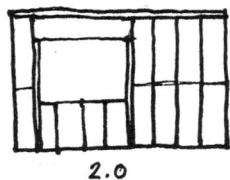

FIGURE 3.85. Rigidity factor 2.0: Membranes, such as plywood sheets, give more rigidity to the construction.

FIGURE 3.86. Masonry infill panels can act as shear walls.

With the exception of catenary and parabolic arches, the line of thrust of an arch will at some point lie outside the middle third (see Figure 3.87). This will make a masonry arch unstable. Domes made of frames of poles set into the ground are not permanent structures. The ground and the tie-loops resist outward thrust.

To resist tension, the builder must either add additional vertical buttressing or tension cables. If this tension is not resisted successfully, the dome will collapse inward onto the occupants, as the lower outerwall falls outward (Figure 3.88).

Figure 3.89 shows a parabolic arch drawn inside an envelope of tangents. The catenary curve is not a conic section. The thrust line lies within the middle third. A bending moment does not occur. The parabolic arch form is stable provided that the horizontal thrust is resisted at the footing level.

A simple way to resist horizontal thrust is to incline the footing. Often the thrust is so little from a parabolic dome that the friction on the footing is sufficient (Figure 3.90).

Construction of arches, vaults, and domes requires expensive wooden centering to support the loads during the construction period. Furthermore, windows are not easy to install in the sloping sides. Masonry construction for vaults and domes in cool climates is rare for low-cost shelter. An almost insurmountable problem is keeping the rain out at a reasonable cost. See Figure 3.91.

Mediaeval vaults are ceilings, not roofs. Elaborate buttresses are necessary to resist thrusts. Weights are enormous. See Figure 3.92.

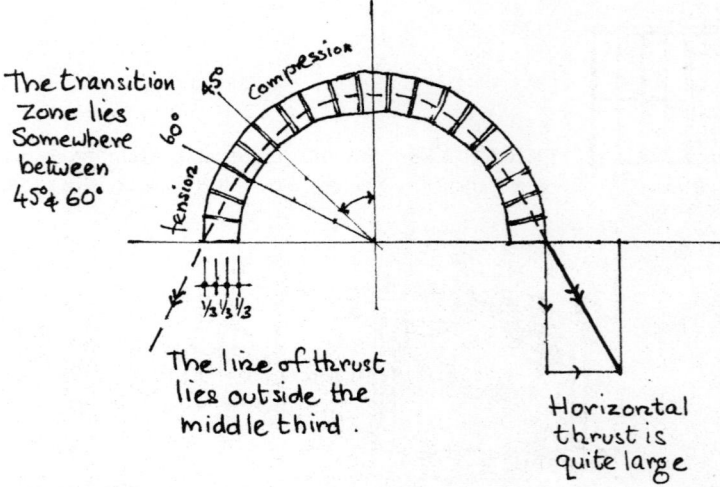

FIGURE 3.87. All masonry arches, vaults, and domes, exert a sideways thrust.

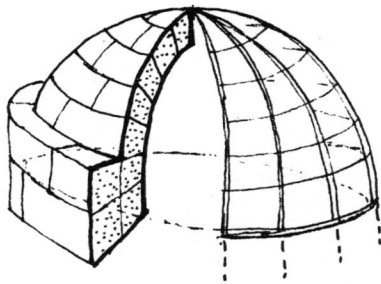

FIGURE 3.88. Two ways of resisting the sideways thrust of domed structures.

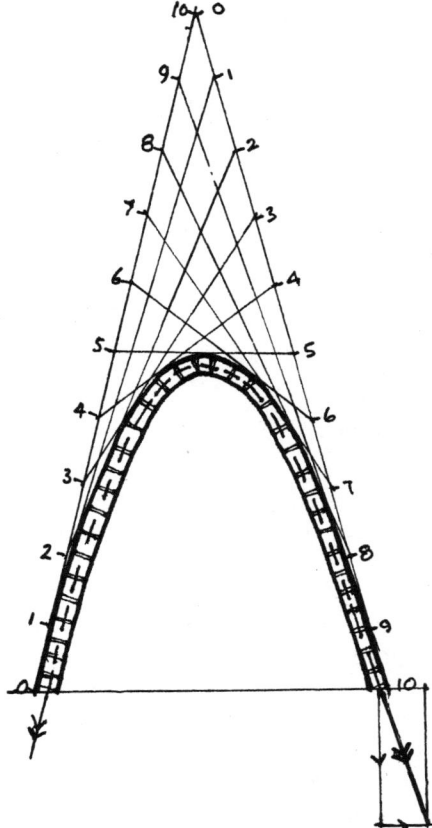

FIGURE 3.89. How to construct a parabolic arch.

73

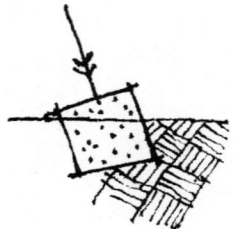

FIGURE 3.90. Inclined footings help to resist lateral movement.

FIGURE 3.91. The keystone is the last stone to be placed in a masonry arch.

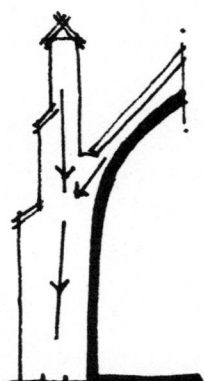

FIGURE 3.92. A mediaeval vault.

The outer dome in Figure 3.93 is made of lead on a wooden framework. A brick cone carries the weight of the lantern. The inner dome is a ceiling. The thrust is taken by iron chains.

In hot climates, light domes of sticks, straw, and other plant fibers provide semipermanent shelter for primitive peoples. When rains come they soon leak and rot. Vermin remains a problem and fire, too. This sort of accommodation is fine when possessions are few. Air circulating freely provides much-needed ventilation. The idea of the dome is attractive. It offers the most for the least. See Figure 3.94.

Egyptian architect Hassan Fathy has produced well-designed domed and vaulted structures using sun-dried bricks and ancient construction methods.

Foundations and Footings 75

Parabolic arches are constructed without the use of centering. The burned clay pot is the universal artifact of the Middle East and has found frequent use in shell structures. See Figure 3.95.

Foundations and Footings

Soils

Whether transmitted by a post and beam system, load-bearing walls, or domes, all loads must eventually be supported by the ground, preferably without undue movement of the building. Even if a suitable soil appears to exist on the surface, what lies underneath may be quite different. It has become good practice to dig test pits to a depth of twice the width of the footing to insure that no unpleasant surprises, such as a weaker plane of

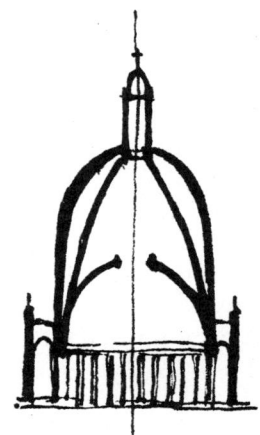

FIGURE 3.93. St. Paul's Cathedral, London England.

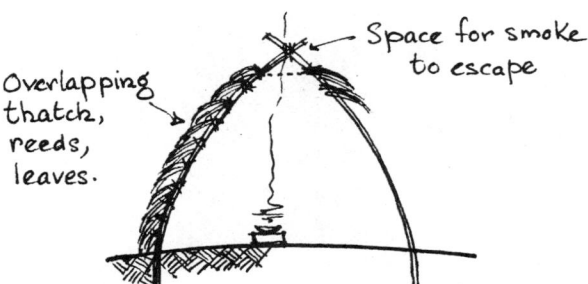

FIGURE 3.94. A simple domed structure.

material, lie just underneath. The absence of this test gave us the famous leaning Tower of Pisa.

Stone and pebbles tend to move up toward the surface, and very fine grained materials tend to move downward. Clays composed of very finely grained materials cling together to form a cohesive mass often found immediately above rock formations. See Figure 3.96.

Uphill water can exert great pressure on clays overlying sloping rock. Sometimes the clay loses its adhesion to the rock and literally flows like mud. This scenario is repeated many thousands of times a year across the world. People remain surprised that the apparently stable hillside is moving, and along with it their home. Incidentally, the hill need not be very steep for this to happen. See Figure 3.97.

Trees are often planted around buildings for ornamental or utilitarian purposes such as wind or sun breaks. Some trees need a great deal of water and what better place to find a humid spot than around building footings? If the footings are sitting on clay, then the roots may dry out the clay and cause the building to settle. If cut down, the tree roots will die and the clay will swell up again. Wood-framed buildings are flexible, but once cracked, masonry walls tend to stay that way. Unless badly fissured or laminated, rock provides an excellent bearing for foundations, and footings are usually omitted. Sand or gravel also provides good bearing, but in Figure 3.98, footings are usually provided to spread the loads. However, it is usually difficult to grow food in rocky or very sandy soil, and human settlements have in the main been formed on alluvial plains along rivers and on the plains of the great clay belts.

Clay provides a good bearing. But when foundations are shallow or the building rests on wooden sleepers or on stone slabs, an uneven settlement may take place that will cause the building to take on a drunken tilt toward the sun. See Figure 3.99.

FIGURE 3.95. Domed structures of the Middle East.

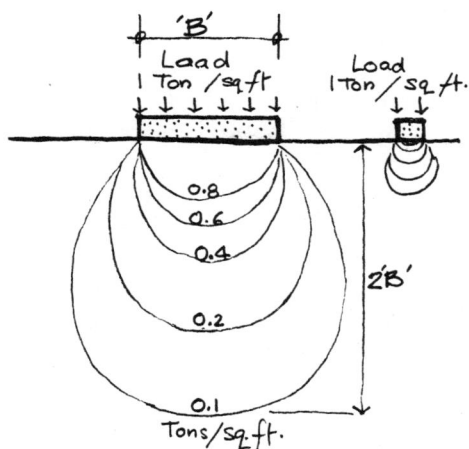

FIGURE 3.96. The pressure on soil under a footing decreases with the depth.

FIGURE 3.97. Landslides caused by water pressure.

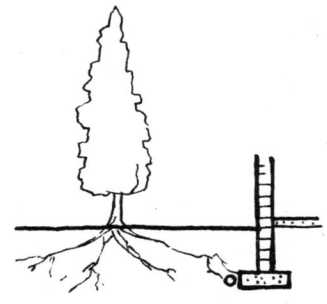

FIGURE 3.98. Tree roots search for water around footing drains.

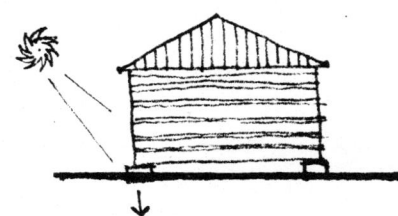

FIGURE 3.99. Problems with shallow footings on clay soils.

This uneven settlement may be caused by the freeze/thaw action in winter or by the shrinkage and swelling of the clay due to sun and rain. Footings on clay are often deep (see Figure 3.100). It should be noted that it was also clay that slid down the hill and also gave trouble with trees.

Spread Footings

To carry loads on moderately strong soils, builders have long resorted to a load-spreading system. This may consist of large stones placed in a trench with smaller stones forming a foundation wall. The wooden floor and subsequent superstructure can then be placed in position. See Figure 3.101.

Spread footings were often built in brick using lime mortar or preferably hydraulic lime mortar, since lime mortar sets up by the absorption of CO^2, a somewhat slow process. See figures 3.102 and 3.103.

When Portland cement became available, a simpler form of construction was possible. Punching shear had to be considered, and rules such as twice the thickness of the wall for width and the 45° angle were devised. See Figure 3.104.

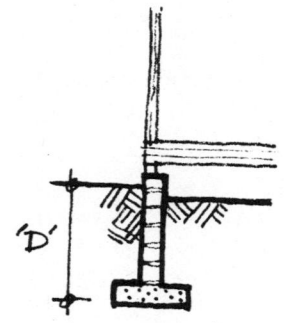

FIGURE 3.100. Clay footings.

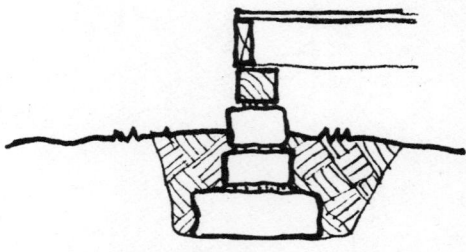

FIGURE 3.101. A simple stone footing.

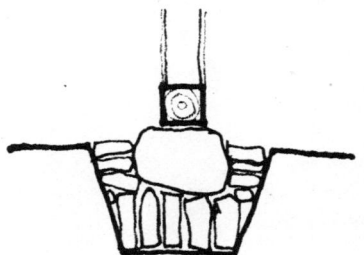

FIGURE 3.102. A traditional Japanese method of keeping floor timbers dry by assuring good drainage.

Foundations and Footings

FIGURE 3.103. A spread brick footing.

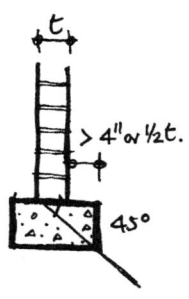

FIGURE 3.104. A concrete footing for light loads.

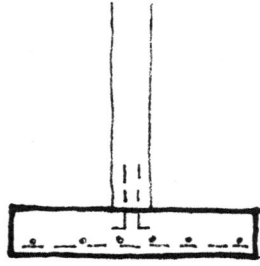

FIGURE 3.105. A reinforced concrete footing for a column load.

For individual column loads, the rules for spread footings are too onerous, and it is usual to reinforce the footings for all but the lightest loads (see Figure 3.105). Heavier loads on plastic soils require deeper footings to avoid simply squeezing out the soil from under the footing.

Although now less common, it should not be forgotten that the gradual widening of the wall near the base will also better distribute wall loads to the soil if separate footings are not used. See Figure 3.106.

Rafts

When they do not wish to dig down for footings, modern builders often float a raft of reinforced concerete on a bed of drained gravel (see Figure 3.107).

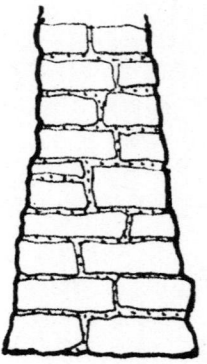

FIGURE 3.106. Gradual widening of the wall near the base.

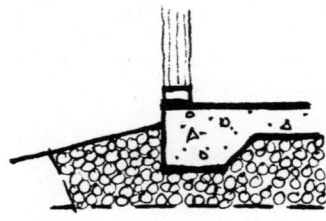

FIGURE 3.107. Floating slab construction.

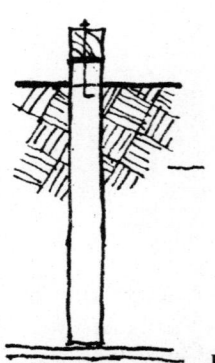

FIGURE 3.108. An end-bearing pile.

Light wooden walls can be constructed on the thickened edge of the concrete slab.

Piles

Sometimes a good bearing is far down into the ground, too deep to economically construct foundation walls. In this case post augers and treated wooden poles can be used as end-bearing piles (see Figure 3.108). Friction piles may also be used—usually they have bulbous ends and are made of concrete. Long before treated wooden piles became possible, it was discovered that wood does not rot if kept permanently wet or submerged. Venice is built on wooden piles.

Stone is used at rising and falling water level marks where the oxygen renewal can cause rot to take place (see Figure 3.109). Piles are often used in marshy and river delta areas.

A dry slab can be constructed using short piles of concrete. The soil is forked over before pouring the slabs. The soil subsequently compacts, leaving an air space. See Figure 3.110).

The Roman method of sinking faggots of wood in marshy areas and then loading them with small stones is well known. (The subsequent drainage of the marshes left the bridges higher than the surrounding ground.) See Figure 3.111.

Buildings and Fire

Fire Spread from the Exterior

Modern fire-fighting equipment and methods can usually prevent the uncontrolled spread of fire in cities. But even so, fire destroys millions of

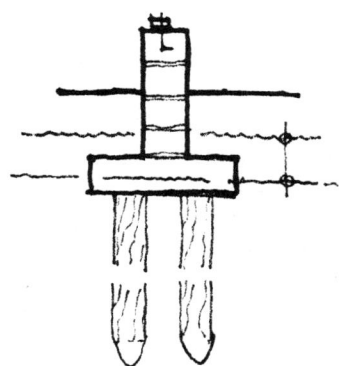

FIGURE 3.109. A stone foundation on wooden piles.

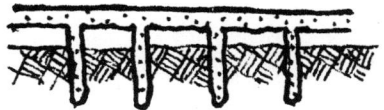

FIGURE 3.110. A floor slab supported by short bore piles.

dollars worth of buildings and contents and claims thousands of lives each year.

Historically, the Great Fire of London in 1666 marked a turning point in the story of man versus fire. Samuel Pepys records London as a "muddled infinity of timber built, pitch coated little houses"—ideal material for a major fire. The London Rebuilding Act of 1667 forbade the construction of timber houses and the construction of overhanging stories. The act also set height limits. In France, King Louis XIV determined that Paris should not suffer a like fate and on August 18, 1667, he signed a Royal Ordinance to the effect that all wooden structures be covered both inside and outside with lath and plaster. Paris never burned, and in fact, Les Invalides, which was built in accordance with this ordinance survived a violent fire that broke out in the dome in 1938. It should not be forgotten that modern Paris sits on a series of abandoned limestone and gypsum quarries, from whence the name for Plaster of Paris. Quarrying continued until Napoleon caused its closing and required the production of plans of the ancient workings, some of which constitute the Paris catacombs.

Masonry walls projected above the roof line as parapets became an accepted architectural detail (Figure 3.112). Fire brands from burning buildings when fanned by high winds present serious risk to adjoining buildings. To reduce the danger, roofs may be required to be covered with slates, clay tiles, cement tiles, or other incombustible materials. Needless to say, flat roofs covered with crushed stone or gravel meet this requirement.

Fire Spread from the Interior

To enhance the likelihood of containment, fire separation walls are frequently carried above the roof line in row or terraced housing in the form of parapets. In masonry and timber construction, to reduce the likelihood of structural collapse, wood joists are fire cut to enable them to fall free if

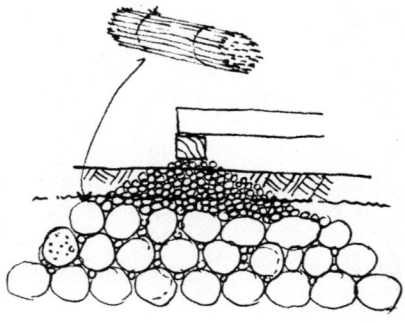

FIGURE 3.111. A foundation built on faggots.

FIGURE 3.112. An accepted architectural detail: Masonry walls projected above the roof line as parapets.

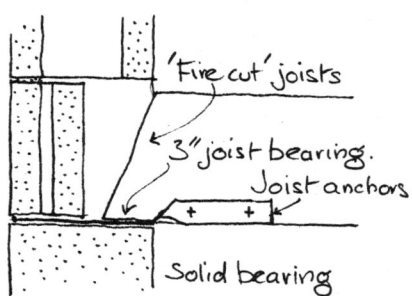

FIGURE 3.113. Fire-cut wooden joist.

the floor is destroyed by fire (Figure 3.113). The masonry walls left standing will probably contain the fire or at least slow down the rate of fire spread.

Combustion Principles

Combustion or burning is a chemical reaction involving oxidation, which produces heat as it proceeds. Burning requires that there be present (1) fuel, (2) oxygen or air, and (3) the proper kindling temperature. Consider a wood frame structure protected by a plaster or plasterboard interior lining. The fuel is the wood and since it is a solid, it must be heated to convert it into the gaseous form necessary for the proper mixing with the oxygen molecules. Wooden 2×4 in. studs are considered as bulk material, being too large for tinder or kindling.

Plaster and Plasterboard

Gypsum plaster contains about 20% of its own weight as water. As the plaster is heated, the moisture held in the pores escapes. Desorption of moisture and all dehydration reactions involve the absorption of heat and retard the flow of heat through the construction. These latent heat effects improve the performance of plaster as a fire-resistant building material. Failure in fact nearly always occurs because the protective lining or membrane becomes detached from the wood frame or the fire gains access to the space from another quarter. Plaster or plaster wallboard also provide protection from rapid flame spread across the surface of lining materials thus giving time for occupants to escape or for a fire department to contain a fire. With the advent of plastics and plastic insulation materials, flame spread and smoke contribution have become very serious problems once more. Exposed to the 1000°F (538°C) of the standard furnace, wood chars at about $1/40$ in./minute, steel fails when unprotected, and even concrete loses 50% of its original strength.

Wooden Construction and Fire Prevention

BALLOON FRAME CONSTRUCTION Balloon frame construction is shown in Figure 3.114.

WALL CONSTRUCTION

1. $5/8$ in. gypsum wallboard.
2. 2×4 in. studs—16 in. oc.
3. $3 5/8$ in. mineral wool—prevents circulation of hot gases.
4. $1/2$ in. plywood.

Finish on exterior and vapor barriers is not indicated. The plasterboard is secured with at least $1 1/2$ in. #13 nails with $7/32$ in. ϕ heads at 6 in. oc.

FLOOR CONSTRUCTION

1. $5/8$ in. T & G flooring.
2. Asbestos paper.
3. $5/8$ in. plywood.
4. 2×8 in. joists.
5. $5/8$ in. gypsum wallboard.

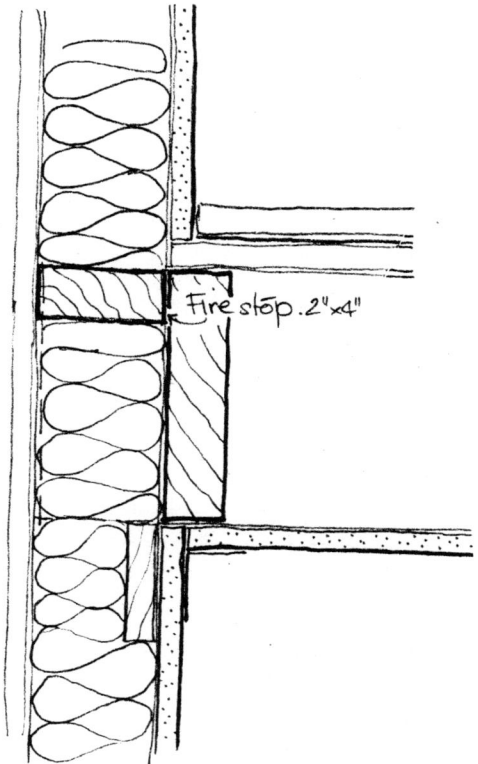

FIGURE 3.114. Balloon frame construction.

Placed between the stud and joist, the fire-stop prevents twisting of joists (see Figure 3.115). A wooden structure composed of solid elements, such as occurs in plank-frame construction, provides few air spaces and slows down the access of the necessary oxygen. Wood-framed structures must always be properly fire-stopped for the same reason, and it should be noted that platform construction automatically provides fire-stopping, while balloon frame construction does not.

PLANK FRAME CONSTRUCTION   Plank frame construction gives better fire protection and is a good system if you have plenty of wood. Moderate insulation is provided by 2 in. of wood. Plank frame of solid wood in-fill is often completed with a cavity brick veneer on the outside. See Figures 3.116., 3.117. Solid planking provides no air spaces for oxygen supply to a fire.

86 Classical Construction Methods for Low-Rise, Low-Cost Housing

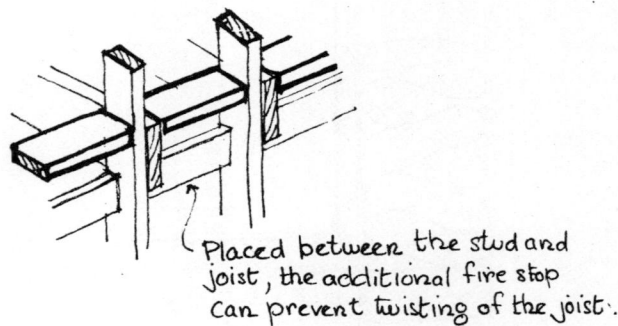

FIGURE 3.115. Fire stopping in balloon frame construction.

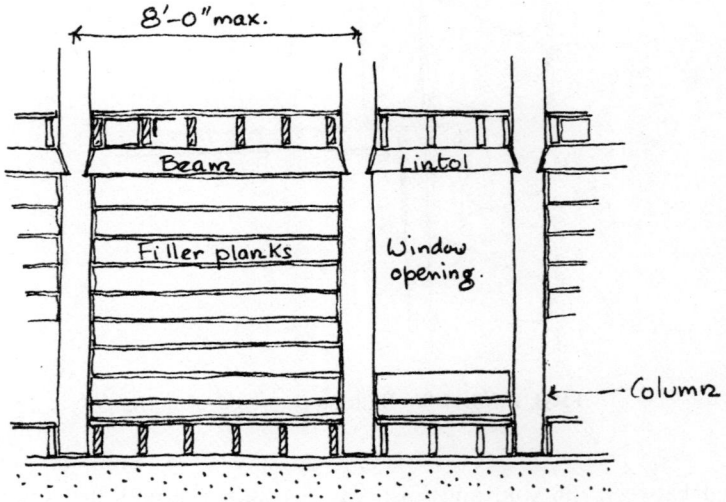

FIGURE 3.116. Plank frame construction.

COMPARISON Figure 3.118 shows post and beam and platform frame constructions compared.

Protection from Water

Thatch Roofs

From the temperate grasslands to the equator, thatch is the primordial universal roofing for permanent buildings. See Figure 3.119.

The roof covering in Figure 3.120 consists of bundles of reeds or straws tied to battens or spars with tarred twine or untreated plant fibers. Bundles are laid with butt ends pointing toward the eaves and are securely tied to battens spaced between 8 and 12 in. on center. Bundles must be tightly packed and are often beaten into position. Twisted rods of pliable willow or bamboo are interlaced through and over the bundles. These rods are usually

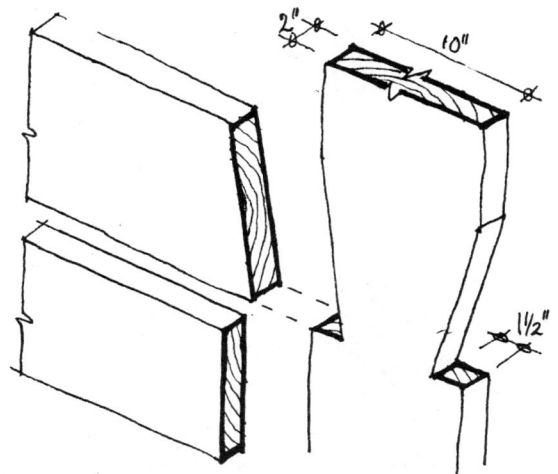

FIGURE 3.117. Lintel support in plank frame construction.

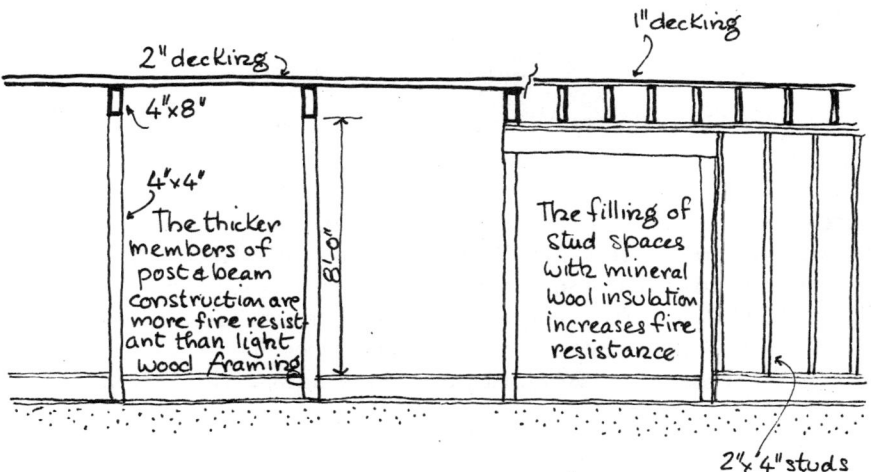

FIGURE 3.118. Elevations of post and beam and platform frame constructions.

FIGURE 3.119. Hut, Chagga, Tanzania.

FIGURE 3.120. Cottage, nineteenth-century England.

fastened to the battens to avoid stripping of the roof in strong winds. The thickness of the thatch varies from 9 to 16 in. The pitch of the roof is at least 45°. Reed thatch when properly maintained will last for 60 years in England. Thatch burns easily and harbors vermin. Most people try to replace it with other materials when possible. Thus it has become a status symbol for the middle class at least in parts of Europe with a strong thatch tradition.

Thatch keeps out rain by expanding when wetted and thus packing tightly together. When dry, air passes through fairly easily, which is why it is so pleasant in hot, humid Tanzania. See Figure 3.121.

Slate, Shingle, and Tile Roofing

SLATES. Slates are a common roofing material in Wales and the Ardenne region of France where they give a characteristic color and roof form. Unless they are very large, slates require a sharp roof pitch. Hips are difficult to form and are found only on middle-class housing, Monotony can be achieved whatever the materials and methods of construction, but industrialization increases the chances. See figures 3.122 and 3.123.

Besides being thin, slates lend themselves to capillary action, which may cause wood rot, particularly if an underfelt is used without battens. See Figure 3.124.

Slates can be center nailed in windy locations. This is an expensive procedure because more slate per square is necessary. See Figure 3.125.

Stone slates, which are usually some form of limestone, come in various sizes and are very heavy. Some can be considered as slabs, having been held on by oak pegs. These slabs being of the same golden stone created a unity between wall and roof materials. Figure 3.126 shows a stone slate.

SHINGLES Figure 3.127 shows 210 lb asphalt coated felt shingles laid on roof sheathing and secured by nails. The shingles measure 3 × 1 ft. Asphalt shingles are economic, easy to lay, light, very efficient, and good for 20 years. If properly installed, they will not easily blow off. Available in many colors, they give an air of sameness to suburban North America. This is possibly because of their lack of thickness. Since about 1890 asphalt shingles have been used in North America. Unlike slate, they do not break easily, and unlike tile, they are not damaged by repeated freeze/thaw cycles.

FIGURE 3.121. Nineteenth-century Japanese thatched roofs showing decorated ridges. Above, a tile ridge; below, a bamboo ridge.

FIGURE 122. Slate roofs, Châteaudun, Northern France.

FIGURE 3.123. Building that dates back to the Industrial Revolution, Neath, Wales.

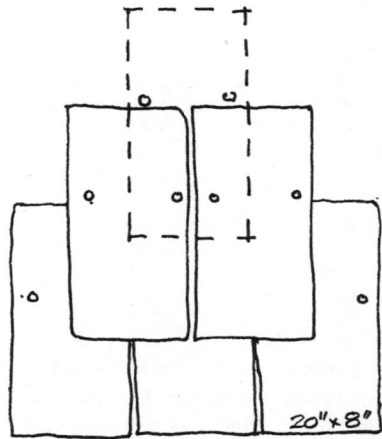

FIGURE 3.124. Center-nailed slates.

Protection from Water 91

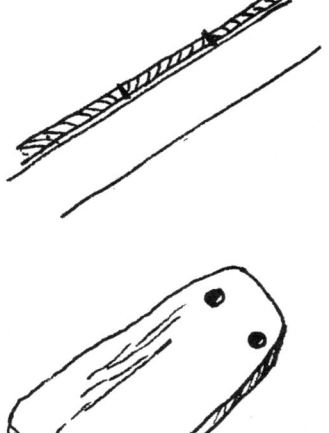

FIGURE 3.125. Close boarding for center-nailed slatework. Battens can be used for head-nailed slates.

FIGURE 3.126. A stone slate, the Cotswolds, England.

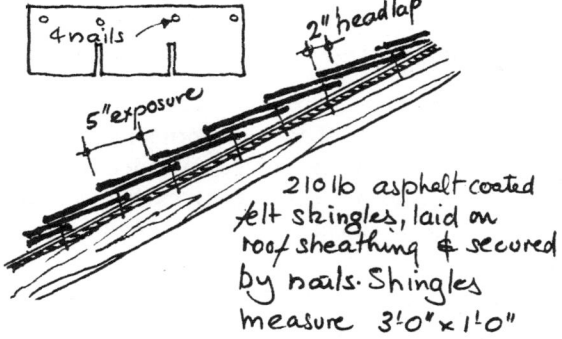

FIGURE 3.127. Asphalt shingle construction.

There are no standard sizes for wooden shingles. Those shown in Figure 3.128 are machine-cut Eastern Cedar $15 \times 7\frac{1}{2} \times \frac{3}{8}$ in., tapered to $\frac{1}{16}$ in. Hand-split Western Red Cedar shingles give a heavily textured roof, but are very expensive. Wooden shingles have been in use for centuries, and wherever used, they are laid with the overlap system used in tile and slatework. Western Red Cedar and Redwood are preferred woods, but almost any wood that can be split has been pressed into use. In nineteenth-century Japan, $7\frac{1}{2} \times 5 \times \frac{1}{8}$ in. shingles were secured to a light, close-boarded roof with bamboo pegs. The pegs were bent over and broken to form a head. The

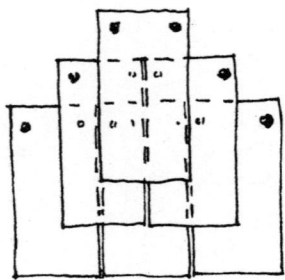

FIGURE 3.128. Head-nailing wooden shingles with large nails.

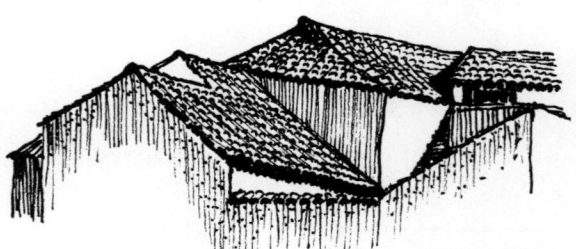

FIGURE 3.129. Tile roofs at Ronda, Spain.

shingles were frequently blown off in windstorms and were extremely dangerous from the point of view of fire spread. Wooden shingles are fine when nonsubterranean termites do not exist. They can be treated against fungi and flame spread. Asbestos cement shingles were popular after the war, but are less so now.

TILES The typical and apparently haphazard arrangement, with sharp color contrast of red tile and white stucco is shown in Figure 3.129. Clay tile roofing has been popular for centuries, and the material provides a characteristic surface texture and a very definite color. It is almost impossible to think of traditional Spanish or Italian housing without clay tile roofing. Similarly, the material is common in Latin America because of European colonization. But it must not be forgotten that the tile of terra cotta is common in Japan, China, Korea, and Singapore on all but the most humble dwellings. In England, the slightly curved tile is called a "pantile" and is used chiefly on barns. The plain tile is used on middle-class housing and has the same problem as slates: a side wind can force rain under the side lap and cause leaks. Interlocking tiles solve this problem. Being bigger, they are also more economic to lay. Concrete tiles are available in almost as many shapes as clay tiles.

Figure 3.130 shows a nineteenth-century slightly curved tile common in China, Japan, Singapore, and Europe. In the example the water runs partly under the overlap. In nineteenth-century Japan, clay tiles were often laid up in beds of mud spread on the roof boards. They relied basically on their self-weight to hold them down in case of windstorms.

The width of the space between tiles prevents water from gaining access by capillarity. Tiles are hung to battens. There is no roof sheathing. Depending on roof pitch and winds, there may be some rain penetration. See figures 3.131 to 3.134.

In Figure 3.135 notice the difference between the slope of the tile and the slope of the rafter. Ribbed surface prevents sideways movement of water which results in less water in the interlocked side joint.

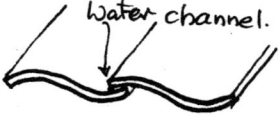

FIGURE 3.130. A nineteenth-century slightly curved tile common in China, Japan, Singapore, and Europe.

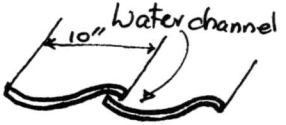

FIGURE 3.131. A twentieth-century clay pantile with the water channel clear of the overlap.

FIGURE 3.132. 950 lb/square Spanish tiles $13\frac{3}{4} \times 9\frac{3}{4}$ in.

FIGURE 3.133. 1250 lb/square Mission tiles 16 x 8 in.

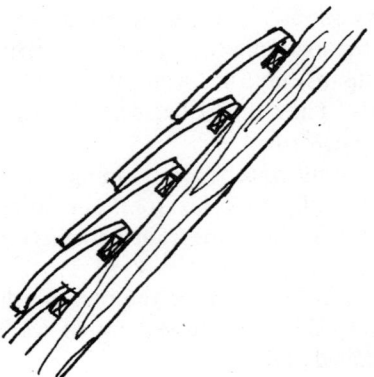

FIGURE 3.134. Plain burnt clay (terra cotta) tiles with hanging ribs.

FIGURE 3.135. Patented interlocking terra cotta or concrete tiles.

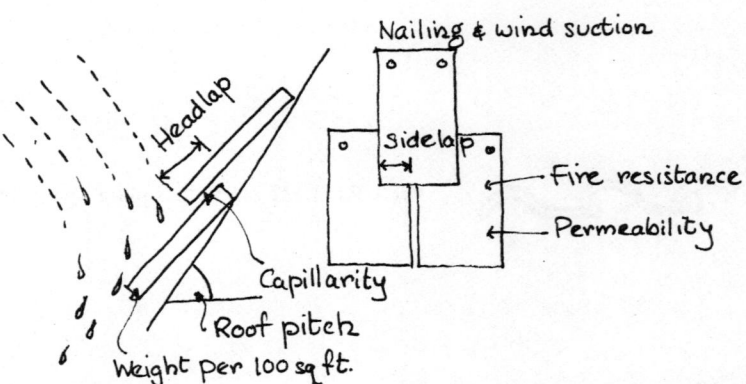

FIGURE 3.136. A summary of the basic requirements for pitched roofs using small overlapping units.

Principles for Sloped Roofs

Sloped roofs are widely used in the construction industry. There are several advantages of their use.

1. Secure nailing against blow offs by wind suction.
2. Use adequate headlaps and sidelaps to reduce water penetration.
3. Avoid water entry by capillarity.
4. Allow for good fire resistance of the units.
5. Make units reasonably impermeable.

Figure 3.136 provides an illustrated summary of these points.

Roofing with Large Sheet Materials

The new universal roofing materials are galvanized corrugated iron or cement asbestos sheets. They are frequently used with concrete block walls and with a low pitch. They can be seen on all continents on new housing developments and are light, fireproof, vermin proof, waterproof, and economic. See Figure 3.137.

Flat Roofs

In environments where there is virtually no rain, a building system of this sort can be used: the roof can be constructed of mud and finished with a reasonably impervious finish such as lime wash mixed with cow dung. The roof is laid to gentle falls and is provided with a projecting scupper to throw the water clear of the walls and thus prevent erosion. As a refinement, metal containers can be cut open, lapped together and held down by stones. With better control of materials in terms of quality and the use of more permanent finishes, such a building system can be used to provide good quality shelter. The life span of the houses however depends on regular maintenance. When the roof is required to provide protection from persistent rain—and at the same time be essentially flat—a bituminous covering must be used. See figure 3.138 and 3.139.

FIGURE 3.137. A profile of corrugated galvanized iron or asbestos cement sheets, 8 x 2 ft.

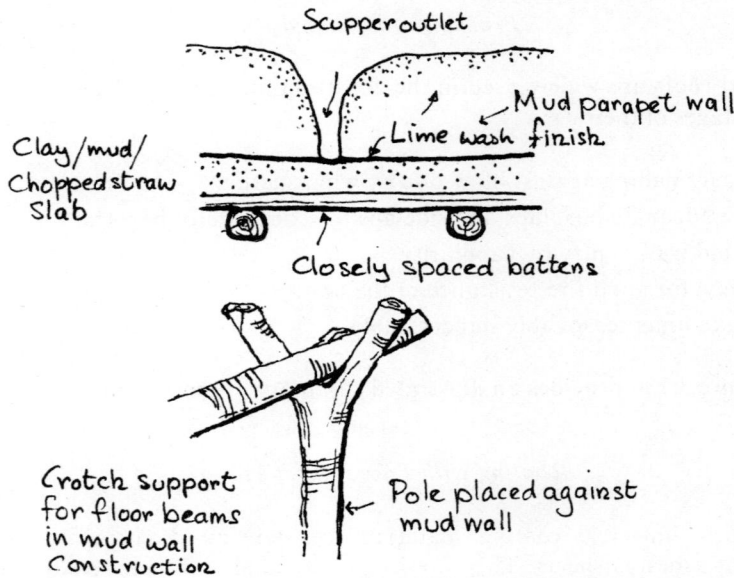

FIGURE 3.138. Roof construction in hot climates with low rainfall.

A slope of 1:50 is necessary for adequate drainage. A basic but often ignored requirement for flat roofs is that they should be sloped to shed water. For a roof to leak there are three essentials: gravity, a hole, and water. Gravity is always with us. Holes occur in the best of construction, but if the roof is well drained, little serious damage may occur. Two systems exist for applying bitumens. The first and very ancient method is the application of natural rock asphalt obtained from "slime pits." This material is now processed with asphalts obtained from petroleum distillation to obtain a product called "mastic asphalt." Mastic asphalt is applied to rigid decks, preferably of reinforced concrete, in two thicknesses for a total of $\frac{3}{4}$ in. Asphalts degrade rapidly in sunlight; but because of the mineral fillers and its thickness, mastic asphalt provides a waterproof surface that can be walked on. Extremes of heat and cold can cause trouble. In the first instance the roof covering will tend to flow, and in the second, it will certainly crack and leak. The second system consists of asphalt from petroleum distillation interlaced with layers of reinforced membranes of compressed wood fiber felts or glass fiber mats. Covered with a layer of crushed stone the system performs moderately well in cold climates and is satisfactory in temperate zones. See Figure 3.140.

The range of temperature at the membrane is from $-30°F$ to $180°F$ in countries with a continental climate. Until quite recently, building insula-

Protection from Water 97

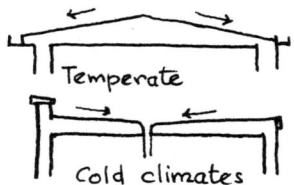

FIGURE 3.139. Roof drainage for temperate and cold climates.

tions were made from materials that absorbed and retained water. The insulation, when wet, lost much of its thermal resistance, and if made from organic materials, would probably rot. Naturally these insulation materials had to be covered by the asphaltic membranes. To stop water vapor from condensing under the cold exterior membrane, a vapor barrier was added, either under the roof deck or under the roofing insulation. Subjected to a temperature range of up to 210°F, this sealed spaced would blow up like a balloon; but unlike a balloon it did not deflate; so the asphaltic membranes cracked and leaked. See Figure 3.141.

The range of temperature at the membrane is less than 20°F. With the arrival of plastic foam insulations, which are impervious to water, the insulation can be placed on top of the membrane, which now has to undergo a very small temperature change. The vapor barrier is not required. The roof must be drained at membrane level and at the top of the insulation. Theoretically and in reality, the protected membrane roof works well, but unless wood and other roof coverings are in short supply, membrane-covered flat roofing systems are at present more expensive than sloped-roof systems for low-cost single- and small multifamily dwellings.

Walls

If there is water and a hole and a force to move the water across the hole, then leaks will develop. Inward sloping cracks would move the water faster toward the interior. If there is not too much water at any one time and the wall is thick enough, masonry—and to a degree, wooden walls—may be counted on to act as temporary reservoirs. Masonry walls are full of

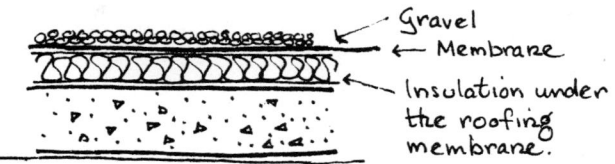

FIGURE 3.140. Conventional flat or low-slope roof construction.

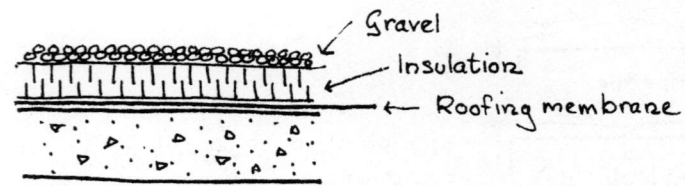

FIGURE 3.141. A protected membrane flat or low-slope roof construction.

capillary tubes, if not outright cracks. Dense stones and hard-burned bricks concentrate water at the weakest link, the joints. The use of high-strength cement mortars does not help because of their greater shrinkage. See Figure 3.142.

Various coatings and washes have been tried over the centuries. In Figure 3.143 mortar coatings (shown in black), which are usually called "stuccos," "renderings," or "exterior plaster," are based on the exclusion principle. They work quite well if the wall does not move or "work" and if the stucco is made of slow-setting lime mortar. Annual coatings of lime wash freshen up the surfaces and seal small cracks. Latex paints perform much better, but another modern material, silicone, has not proved to be as effective as hoped.

Shown in Figure 3.144 is a cavity wall that provides a more or less complete break in the wall. Water that may cross the outer wythe is drained back to the outside. The weep holes also provide pressure equalization for the cavity. The same pressure P_o exists across the outer wythe. Therefore, water movement by the force of pressure differential is eliminated. The effectiveness of the system is greatly helped if an interior air barrier separates P_i from P_o. The outer wythe of the cavity wall can be considered as a rain screen. The screen principle can be developed for outer forms of construction. A tile roof is a rain screen.

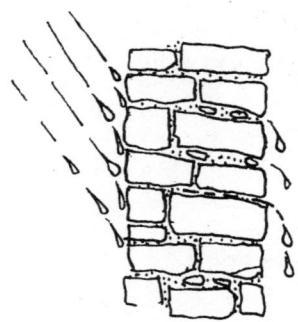

FIGURE 3.142. Water penetration of a masonry wall.

Protection from Water 99

FIGURE 3.143. Surface protection of a masonry wall.

In Europe, hollow terra cotta blocks are common. They are modified cavity walls. It is not uncommon to see the "weather wall" of old masonry buildings protected by boards or metal sheeting. The use of timber planks as a rain screen is part of our prescientific building experiments. See Figure 3.145.

Walls of straw or reed thatch are rain screens (see Figure 3.146). Little or no pressure differential is possible, and the wind soon loses its ability to force the water droplets across so many kinetic energy barriers. Ship lap horizontal planks are a rain screen in wooden construction, just as tile hanging is on masonry walls. See Figure 3.147.

It is all too easy to solve one problem and create two new ones. In the case of the vertical overlapped joints, the close contact between the back of the plank and the solid sheathing creates ideal conditions for rotting. In Figure 3.148 the planks are fastened to horizontal battens to one side only. Unless the battens are stout, it is practically impossible to nail to them, but the circulation of air is good. If solid sheathing is placed under the battens; then the cost mounts. The cover joints keep most of the water out of the wall, and care must be taken to fasten the plank on one side only; otherwise,

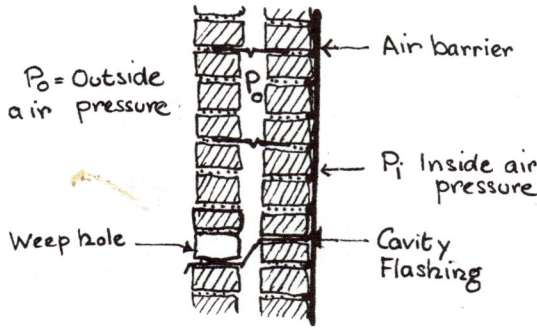

FIGURE 3.144. A pressure equalized cavity wall.

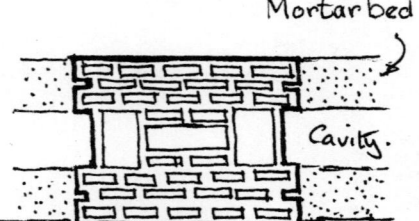

FIGURE 3.145. A British V-brick developed from Continental terra cotta blocks.

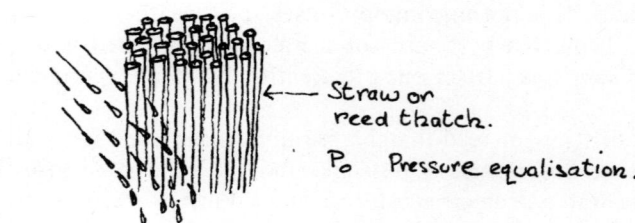

FIGURE 3.146. A straw or reed thatch wall.

FIGURE 3.147. A wood stud wall.

FIGURE 3.148. Overlap joints are used to reduce water penetration on vertical planking.

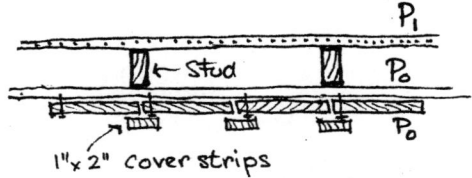

FIGURE 3.149. Horizontal supporting battens provide air circulation behind the planks.

the plank might split. One conclusion might be to use horizontal planking as being more logical (see Figure 3.149).

Joints

Figure 3.150 shows a plan with concrete panels about $4\frac{1}{2}$ to 5 in. thick. Joint tolerances are important. Large panels concentrate most of their problems at the joints. Failure to apply the principles of the rain screen wall has lead to some spectacular failures. It can be said that, no sealant yet invented will seal joints in a single application, and for the life of the building.

Protection against Cold

The Theory of Insulation

Resistance insulation against the cold consists of the creation of air spaces and air films. Since still air offers the most effective insulation readily available, means must be found to prevent air movement in the air spaces

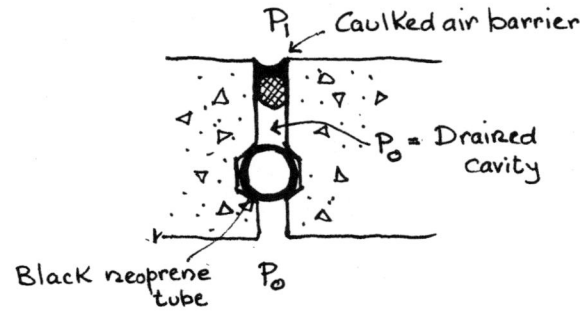

FIGURE 3.150. A joint in a reinforced concrete panel.

created. Light, porous materials such as wood offer greater resistance to heat transfer than dense materials such as stone because air is trapped in the pores of the wood. Materials with a greater resistance (R) to heat transfer than that given by wood are called "insulating materials." Heat is also lost by air leaking from the building through cracks. As outside temperatures drop, substantial air pressure differences exist across the wall. Because of rot and fire hazards, wealthier citizens often opted for stone walls even in areas where wood was plentiful. Extremes of heat and cold in dry climates both resulted in thick masonry walls. (Very cold winters are dry climates in a very real way). In measurement, 10 in. of fluffy snow equals about 1 in. of water. The entrained air makes snow a reasonably good insulating material, and in early Norwegian dwellings every effort was made to keep the snow on the roof.

Construction in Wood

EARLY FORMS OF CONSTRUCTION COMMON TO MANY COUNTRIES. In "chinking," which is the filling of spaces between squared lumber or log construction, the materials used ranged from lime mortar and oakum to moss jambed between the logs. See Figure 3.151.

Lime mortar or planks were applied to the outside and lath and plaster to the inside. This gave the wall a resistance value $(R) = 3.7$. The wood shrank as it dried, and the lime concrete shrank when setting; so the wall leaked air in quantity and was saved by its plaster air barrier. See Figure 3.152.

MID-NINETEENTH-CENTURY CONSTRUCTION IN NORTH AMERICA With the arrival of balloon framing and thinner walls, insulation became more important. Various materials were tried such as rammed sawdust, seaweed, unburnt bricks. In 1873 mineral wool became available and replaced all other insulations used as infill. Mineral wool was made by blowing oxygen

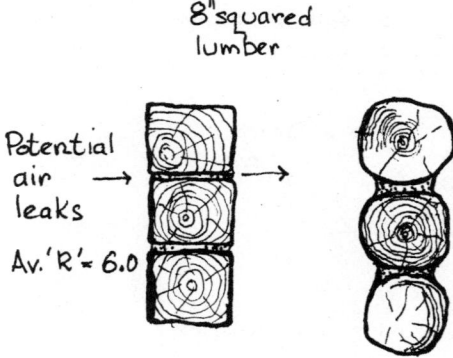

FIGURE 3.151. Early forms of construction common to many countries.

FIGURE 3.152. Spaces between vertical wooden posts were filled with stones and lime concrete.

through the furnace slag, which was an industrial waste from the burgeoning steel industry. See Figure 3.153.

Construction in Masonry

EARLY EIGHTEENTH CENTURY, STONE Masonry construction was generally as thick as possible (see Figure 3.154). Joints were carefully made to avoid air leaks and draughts. Wooden battens were attached to the walls, and wooden laths and plaster applied to the inside. Poorer citizens and settlers would have to make do without the refinement of plaster. For the thermal resistance of the wall with plaster $(R) = 3.5$, and without plaster $(R) = 2.5$.

MID-NINETEENTH CENTURY, BRICK In moderate climates such as Western Europe or Northeastern United States, solid walls of brick with a plastered

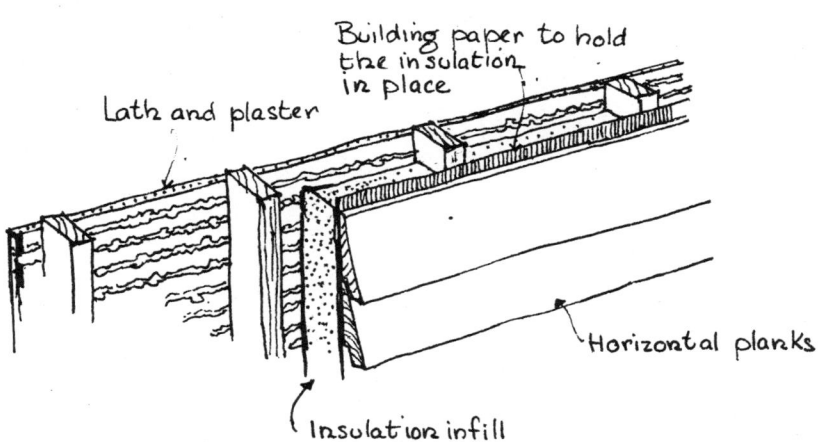

FIGURE 3.153. An insulated balloon frame wood wall.

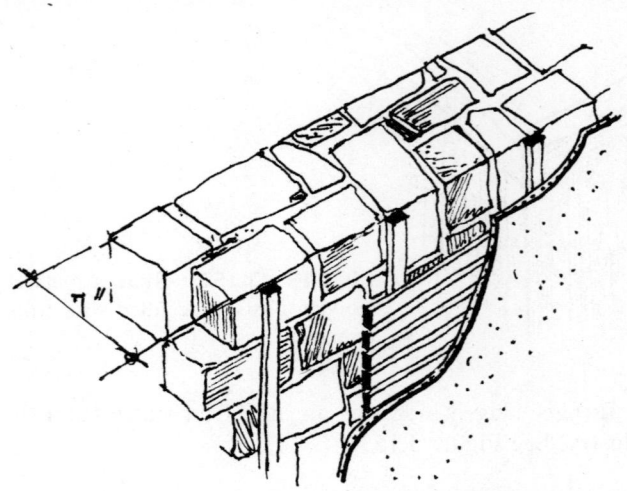

FIGURE 3.154. A solid stone wall with plaster interior.

interior supplied sufficient protection. Bricks were quite costly and well-burnt bricks could only be used on the outside. Under and overburned bricks, which resulted from early brick kilns, were used as back-up masonry. Lime mortar sets up slowly and is not very strong; two good reasons for keeping the joints thin (and more difficult to build). A one-brick wall would not permit the use of back-up bricks in quantity. See Figure 3.155.

MID-TWENTIETH CENTURY On masonry walls, quite logically, the insulation was placed on the inside of the wall. (See Figure 3.156.) This meant that the outside of the wall was colder than before. The larger thermal movements meant more and larger cracks. With the advent of modern plumbing and higher indoor temperatures, many more pounds of water were added to the

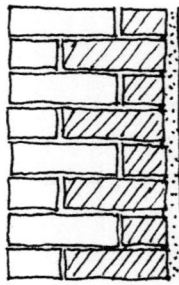

FIGURE 3.155. A one and a half brick wall.

Protection against Cold 105

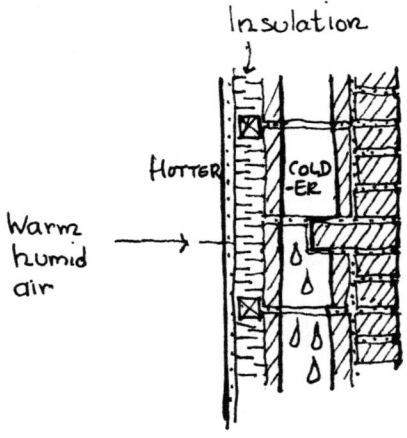

FIGURE 3.156. A concrete block and brick wall.

air inside the house, and when this air passed out through the wall toward the cold, it condensed. Whether made of masonry, concrete, or wood, water inside a wall construction is not a good idea.

VAPOR BARRIERS   The idea of a vapor barrier, shown in Figure 3.157, came about as a means to stop the air diffusing across the wall and with the air, of course, the water vapor in the air. But *diffusion* means "the spontaneous molecular interpenetration of two fluids without chemical combination and usually at the same temperature." This is not exactly what happens. Much more water vapor is carried into walls by plain old-fashioned air leaks, the same air movement that caused so much trouble in the eighteenth century. The vapor barrier is still a valid idea, but it is really an air barrier. Polyethylene sheet and asphalt-coated papers are two common air barriers (vapor barriers). And to stop air leaks around openings of other junctions and interfaces we need the "elastomers," the artificial rubber caulkings.

LATE TWENTIETH CENTURY   A fairly logical solution to insulated masonry walls has now been developed under the general title of "rain screen walls, an example of which is presented in Figure 3.158. The insulation is placed

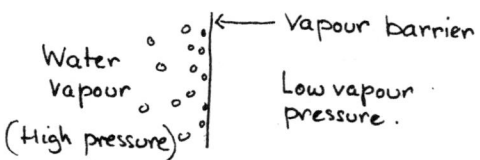

FIGURE 3.157.   The vapor barrier.

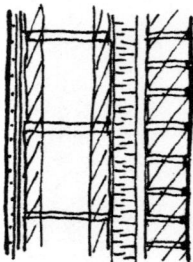

FIGURE 3.158. A rainscreen wall.

on the outside of the inner load-bearing block wall, thus eliminating thermal movements. A cavity and a rain screen exterior of brick complete the wall. This is a costly wall and cannot compete with the insulated wood frame dwelling for low-cost housing in cold climates.

Protection from Heat and Humidity

Hot and Dry Climates

THICK WALLS   Hot dry climates have large daily temperature differences because of the heat loss at night by radiation to the clear sky. Traditionally, walls are thick to take advantage of the principle of thermal inertia whereby the walls heat up slowly in the day and remain warm for the cool nights. The interior temperature is thus maintained within an acceptable range. Plans tend to be square. Buildings share walls. Courtyards provide shaded areas. Plants and awnings provide shade. White surfaces absorb less heat. See figures 3.159 and 3.160.

Adobe was not the only material used. Figure 3.161 shows a multistory building in a Pueblo village in Arizona. The Pueblo people began using stone for building shortly after the time of Christ. The walls have a stucco or rendering on the exterior. The roof construction is outlined under flat roofs.

EVAPORATIVE COOLING   Evaporation decreases the dry bulb temperature. Energy for evaporation comes from air in the room. Vegetation helps in evaporative cooling. Sometimes a bowl of water is placed at the bottom of air shafts and thus cools incoming air. See Figure 3.162.

The climate in Iraq is hot and dry. Ventilation is provided to the interior by roof-top air scoops. The scoops face the prevailing wind. As the air descends, it is cooled by conduction as it comes into contact with the cool party wall. See Figure 3.163.

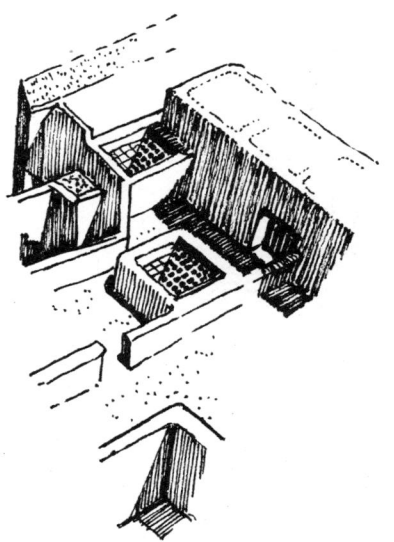

FIGURE 3.159. Roof terraces, Dar Chahed, Tunis.

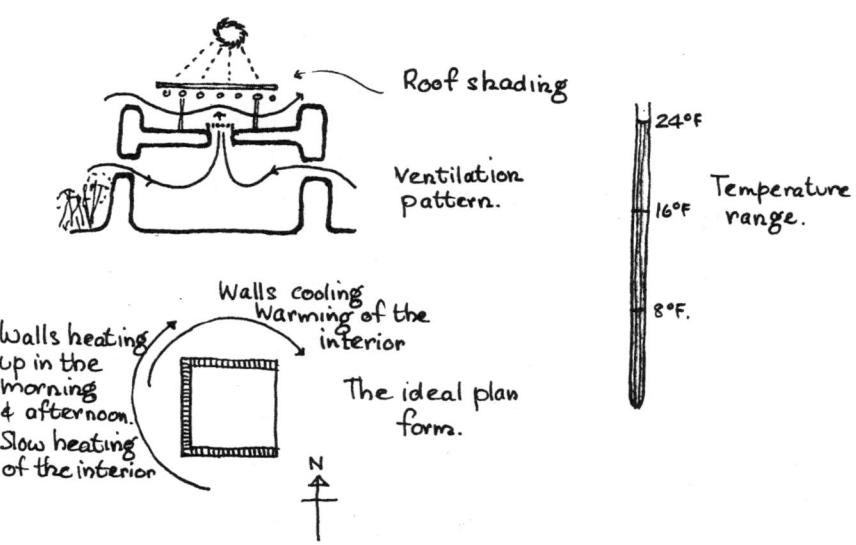

FIGURE 3.160. Thick walls, shading, and an ideal plan shape.

107

FIGURE 3.161. Arizona, USA: A multistory dwelling in a Pueblo village.

Hot and Humid Climates

THIN WALLS Figure 3.164 shows a bungalow of the type built by pioneering European families for a hot and dry climate. Unless repeatedly painted with lime wash or paint, corrugated galvanized iron sheets or asbestos sheets are poor roofing materials because they have no thermal lag.

Hot humid climates have only small daily temperature differences—about 8°F. The high humidty produces high vapor pressures that cause people to describe the weather as "close and heavy." Air movement reduces the vapor pressure, but usually quite high air speeds are necessary. Air movement is in fact quite light except when storms strike. Buildings are frequently raised on stilts to take advantage of any breeze. Roofs require a good overhang to throw heavy rainfall clear of walls. Double roofs have a good insulating value against the sun's rays. The building plan is better if elongated to permit good cross ventilation. Mechanical air conditioning is of great value, but depends on a sound industrial economic base. Evaporative cooling cannot work because the wet bulb temperature is close to the dry bulb temperature. See Figure 3.165.

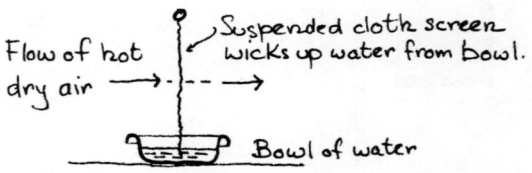

FIGURE 3.162. Evaporative cooling.

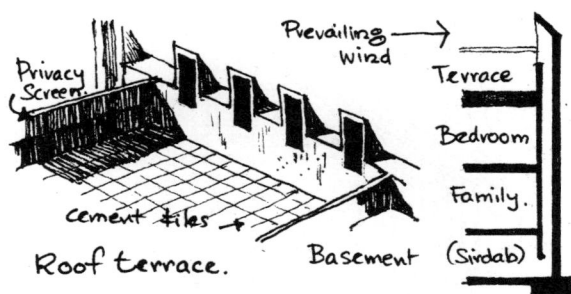

FIGURE 3.163. Oriental courtyard house in Iraq showing a roof terrace with ventilation air scoops.

FIGURE 3.164. Thin walls and shading, Central Australia.

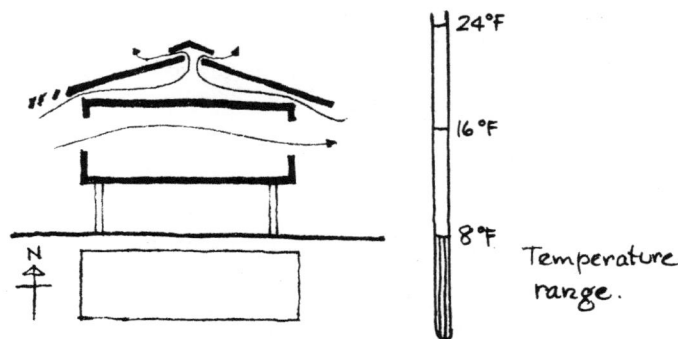

FIGURE 3.165. Air circulation and an elongated plan.

110   *Classical Construction Methods for Low-Rise, Low-Cost Housing*

FIGURE 3.166.   A pole structure, Madagascar.

Figure 3.166 presents a light airy structure sitting on a forest of poles, the use of which permits inspection of the structure for decay and insect attack. Another light porous structure is in Figure 3.167.

Classical Construction Methods

Construction in Wood

LOG CONSTRUCTION   Once a common fuel, wood has become a scarce commodity in many parts of the world. Log construction is slow, uses a lot of material for the volume enclosed, and soon rots. See Figure 3.168.

Figure 3.168a shows 12 in. stovewood walls made of 4 to 6 in. diameter × 12 in. long cedar logs placed between vertical posts, use the up ends of logs. The transformation of woodland to farmland has been a prominent theme in the history of land settlement and land use.

Log construction has been used throughout temperate and cool climatic regions. See Figures 3.169., and 3.170.

SQUARED LUMBER   is an effective way to use lumber for the construction of walls. Figure 3.171 provides an example of a squared lumber construction.

BALLOON FRAME CONSTRUCTION, NINETEENTH CENTURY   The system is not much used now because it is much slower to construct than platform framing. The long 2-story wall lumber nevertheless has less end grain shrinkage than the multitude of plates in 2-storied platform framing. See figures 3.172 and 3.173, and 3.174.

Classical Construction Methods 111

FIGURE 3.167. Mexico: This light porous wall construction encourages air movement.

PLATFORM CONSTRUCTION, TWENTIETH CENTURY   As soon as the foundation walls are constructed, the ground floor can be framed up to provide a flat clean working surface for the construction of the walls. Rapid backfilling will permit easy access for trucks delivering materials. See Figure 3.175.

Platform construction as in Figure 3.176 lends itself to onsite and offsite prefabrication. The lumber is commonly precut to lengths before delivery to the site.

When termites are present, metal termite shields must be used, supplemented with periodic soil poisoning. Termites are spreading northward and

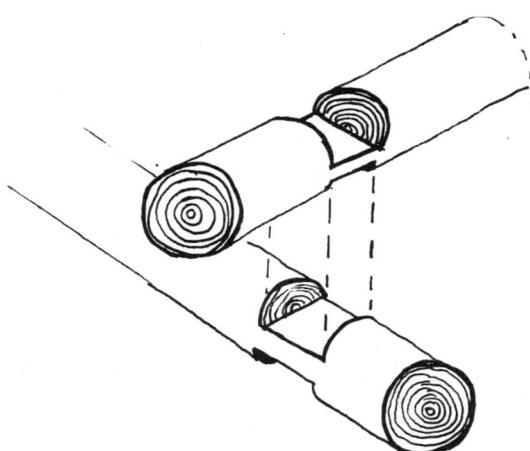

FIGURE 3.168. Log construction: A compression joint designed to resist lateral movement.

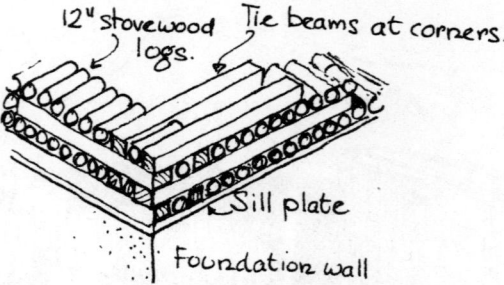

FIGURE 3.168a. A stovewood wall.

the insects are becoming more resistant to insecticides because of rapid mutations. See Figure 3.177.

PRESSURE-TREATED WOODEN FOUNDATION WALLS In cool climates the increasing cost of energy has forced constructors to insulate the previously uninsulated basement walls. Preservative-treated wood foundations can be insulated with economic glass fiber or mineral wool insulation. See Figure 3.178.

Noncombustible Construction

Poured in place reinforced concrete slabs are economic where formwork is repetitive and lumber is expensive. See Figures 3.179 and 3.180. Reinforced concrete and steel constructions are options open to countries with an industrial base. Their use is more and more necessary as construction lumber becomes rare and expensive. As populations concentrate in cities in

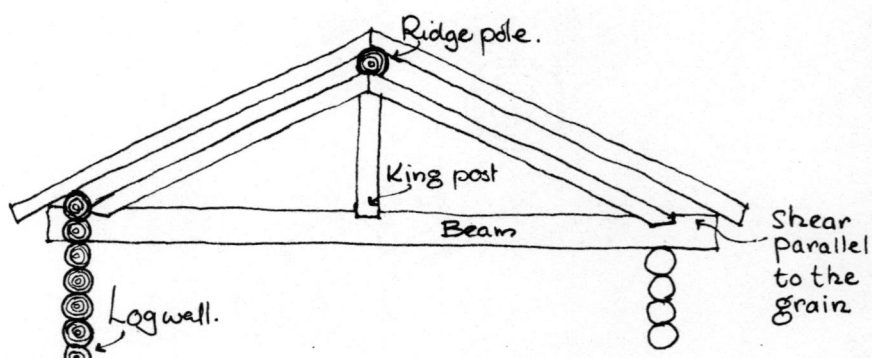

FIGURE 3.169. A roof frame of log construction.

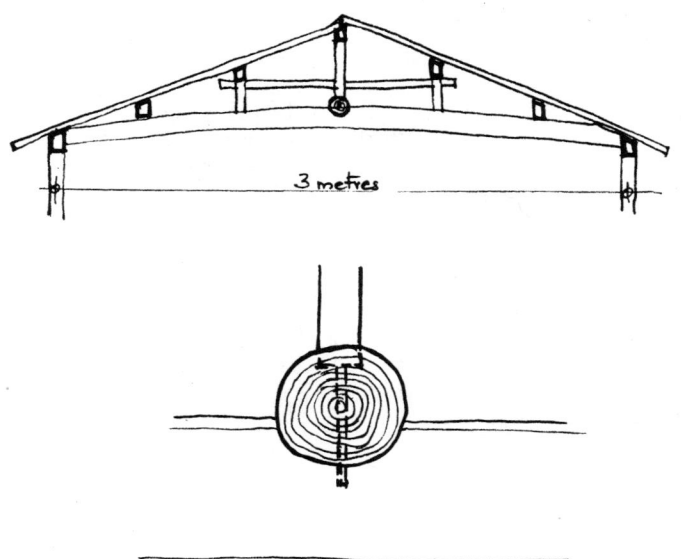

FIGURE 3.170. Example of a roof frame from nineteenth-century Japan.

FIGURE 3.171. Squared lumber: A dove-tailed compression joint designed to resist lateral movement.

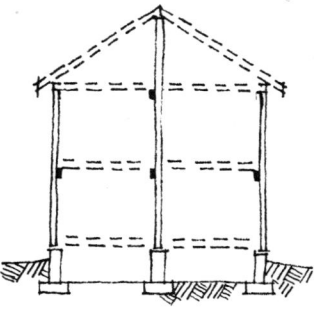

FIGURE 3.172. Balloon frame construction, nineteenth century.

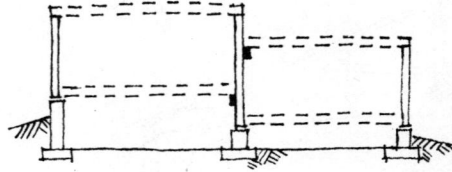

FIGURE 3.173. Balloon framing is more commonly used for split-level construction.

apartments, as opposed to discrete housing, fire resistive construction becomes more necessary. See Figure 3.181.

For low-rise buildings, load-bearing walls either in brick, masonry, or concrete blockwork are more commonly used than post and beam construction of steel or concrete. The masonry walls can be reinforced and tied to horizontal elements when earthquake loads are expected. See Figure 3.182.

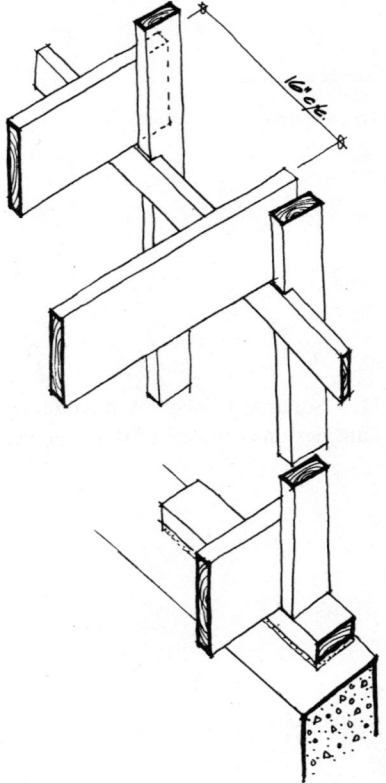

FIGURE 3.174. Isometric drawing of balloon frame construction.

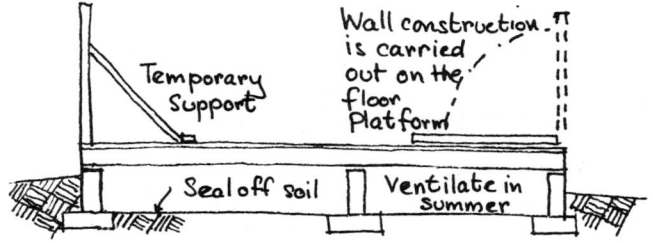

FIGURE 3.175. The ease of building in platform construction.

Roof Terraces

The roof insulation and the wall insulation provide a protective thermal envelope for the roofing membrane and the structural wall. Membranes made of elastomers and asphalt reinforced with polyesters are examples of the constant evolution of materials. Thus the demarcation line between classical and industrialized construction becomes less clear as time goes on. See Figure 3.183.

Cooking, Heating, Sanitation, and Natural Light

Cooking

Cooking requirements are a function of the socioeconomic order. Some years ago, before the recent energy crisis, cooking by the use of solar ovens was promoted by the Indian government. Large numbers of solar ovens were constructed. Unfortunately, Indian peasants eat their cooked meals in

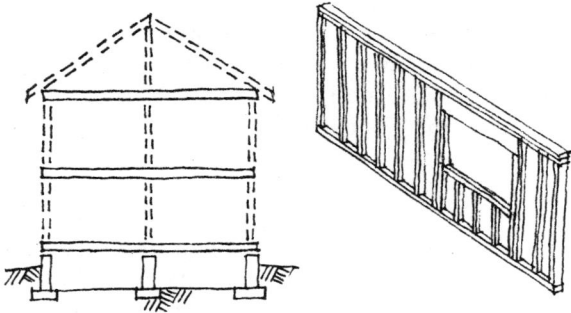

FIGURE 3.176. Platform construction, twentieth century.

116 *Classical Construction Methods for Low-Rise, Low-Cost Housing*

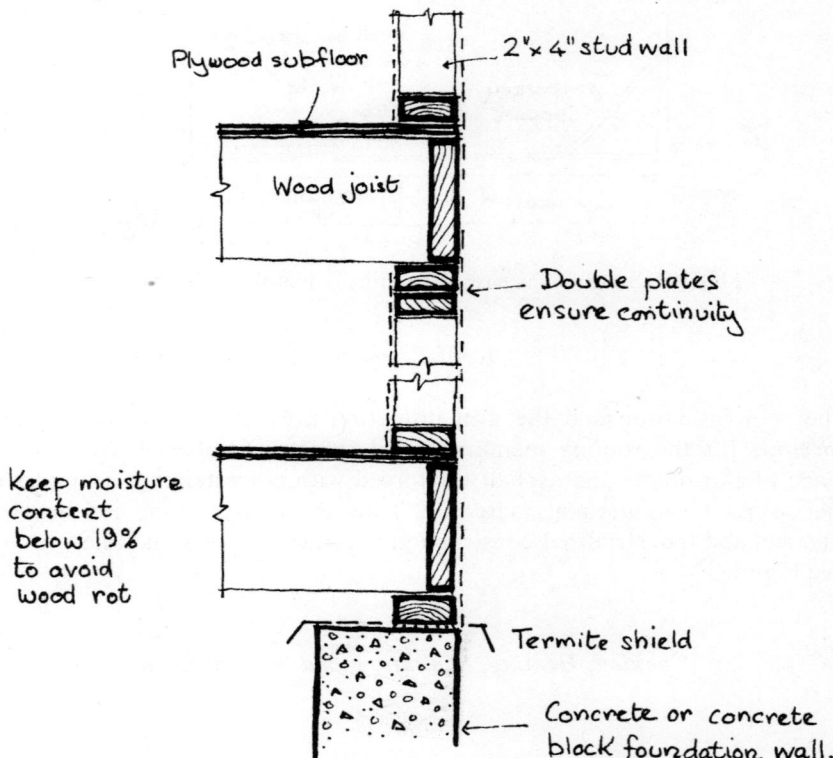

FIGURE 3.177. Section through a wall in a platform construction.

the evening when the sun goes down. Hence solar ovens were not of great use to them.

The hibachi was often made of wood and contained metal or earthenware pots filled with charcoal for handwarming and tea making (see Figure 3.184). The small box in front is a "tabako-bon," a hand spitoon. Japanese raised floor dwellings were derived from Ise Shrine prototypes and were cold in winter—hence the hibachi.

Heating

The tyranny of the wood burning stove that demanded refueling every 6 hours was solved by several ingenious inventors who devised stoves that burned slowly all night. Until the mid-twentieth century however, it was either the wood yard or the coal-scuttle. See Figure 3.185.

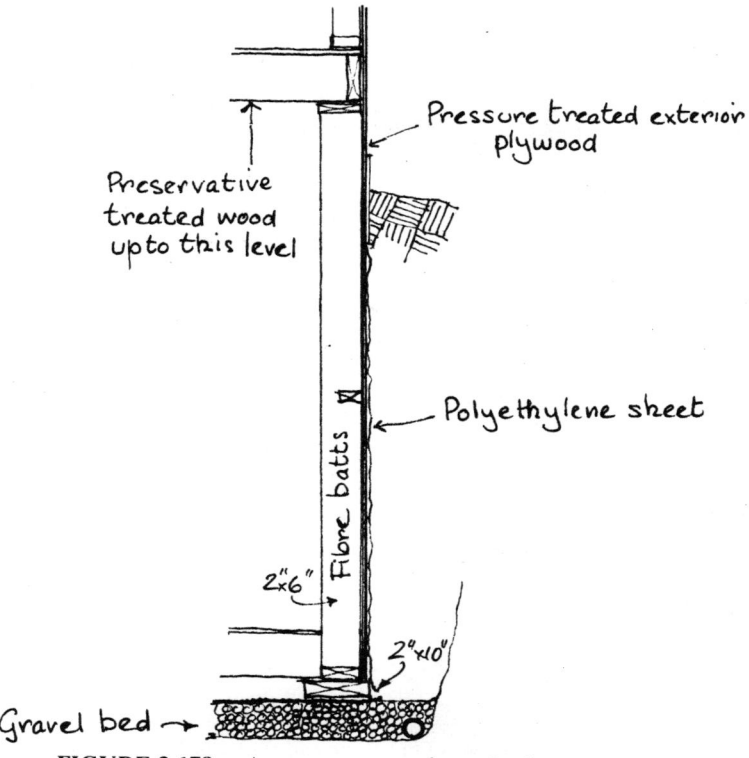

FIGURE 3.178. A pressure-treated wooden basement wall.

FIGURE 3.179. Block of apartments with yellow brick walls and asbestos cement roofing. The human scale is maintained. Architect: Arne Jacobsen, 1952.

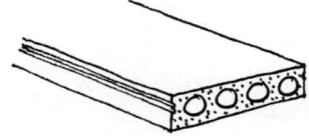

FIGURE 3.180. Precast concrete. Hollow core plank floor slab. Example: 24 in. wide and 6 in. thick for an 18 ft span.

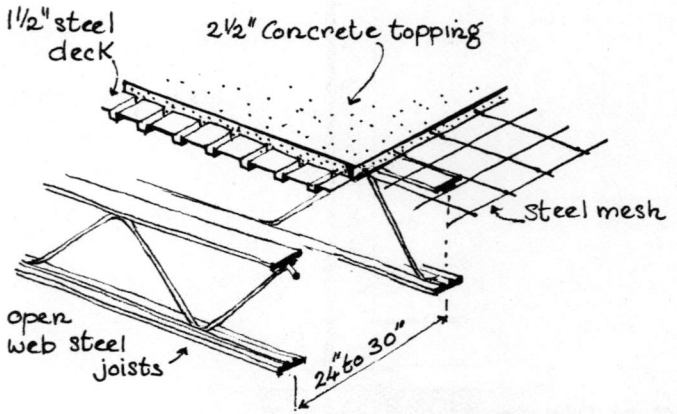

FIGURE 3.181. Lightweight floor and roof construction in steel.

FIGURE 3.182. Block of apartments, Peru. Architect: Santiago Agurto Calvo, 1950.

Cooking, Heating, Sanitation, and Natural Light 119

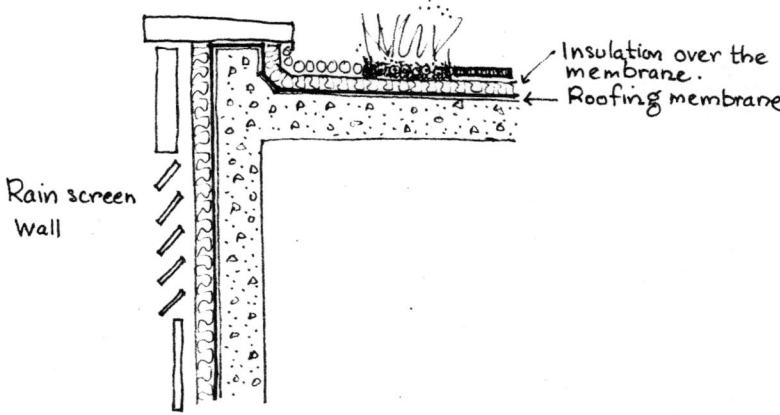

FIGURE 3.183. A roof terrace showing a choice of several materials to protect the sealed cell roof insulation and the wall insulation.

Sanitation

For most of Western Europe and North America, clean drinking water from taps was available by World War I. However, clean drinking water is still not available to much of the world's population. Fresh water is in short supply in much of the inhabited world. Industry also needs vast quantities of water. To make 1 ton of steel it takes 44,000 gal, and 7 times as much to make a ton of aluminum. In cities, nineteenth-century building construction could offer little more than the earth closet or a night collection cart. The 5 gal flush presently is the preferred solution (see Figure 3.186). Many modern communities flush their sewage straight into the nearest river, leaving those downstream with the problem of polluted drinking water.

FIGURE 3.184. Hibachi arranged for company: Japan 1880.

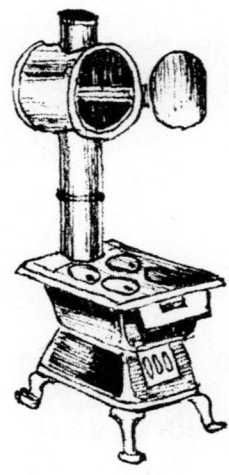

FIGURE 3.185. Combination laundry, cooking, and heating stove, with stove pipe baking oven. Burns coal, coke, wood, corn cobs, and so on. Sears Roebuck & Co. 1927.

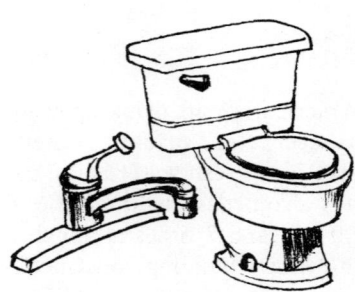

FIGURE 3.186. The 5 gal W.C. and a modern faucet.

FIGURE 3.187. The high narrow windows of eighteenth-century housing permitted natural light to penetrate far into the rooms.

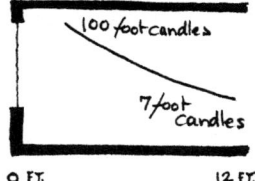

FIGURE 3.188. The intensity of natural light in rooms with high windows.

Natural Light

Glass was known to the Romans, but country cottages in nineteenth-century England often had no glass in their windows. The so-called slums of back-to-back houses were in fact a tremendous improvement over the romantic rural slum.

Before gas and electricity, in North Europe and England, houses were designed to provide the maximum penetration of natural light (see figures 3.187 and 3.188). Unfortunately, much of what has been shown concerns the well-to-do. This is particularly true in terms of what are now known as mechanical and electrical services.

Conclusions

This listing records some of the facts relating to construction methods that have come to light over a period of time. (See also Figure 3.189.)

1. Traditional building methods developed as a synthesis of a social order, the available technology, and simple materials. Climate was usually, but not always respected.
2. The humble dwellings of the poor have not stood the test of time.
3. In the last hundred years, the discovery of steel and Portland cement and the arrival of cheap glass have revolutionized the building industry.
4. In the last 25 years, plastics have brought further changes.
5. Technology can solve the problem of providing dwelling units.
6. Technology alone can never provide homes or shelters without modifications due to social and psychological factors.
7. Traditional building permits whims and fancies; the individual can express himself or herself.

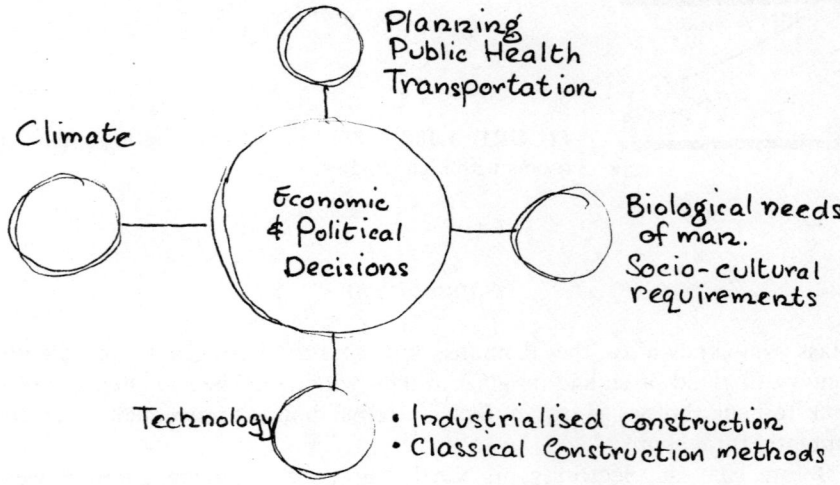

FIGURE 3.189. A systems chart to illustrate the factors related to housing production.

8. To provide the necessary quantity of low-cost shelter, classical construction methods, industrial building systems, and even alternative housing will be necessary.

The final choice regarding construction methods will be made, however imperfectly, in accordance with the political and economic realities of the day. One hopes that the wishes of the people themselves will not be forgotten.

4

Semi-Industrialized Housing Production

Herbert W. Busching

A wide range of industrialization is possible in housing, and this chapter addresses activities that can save housing production time and costs. Semi-industrialization is used in its broadest sense in this chapter and includes onsite fabrication of multiple components and use of preassembled elements. Semi-industrialization can also range to the complete prefabrication of housing units assembled or erected at the site.

Throughout the history of housing there has been a trend to preassemble many of the components of housing. The tendency to industrialize portions of housing production has increased in recent years. This trend increased as transportation routes were improved and as assembly line techniques, refined in the automotive industry, were used in other areas of manufacturing. Certainly, the need for more units, as experienced after World War II, and the high cost of labor-intensive construction served to motivate use of industrialized or semi-industrialized housing. The mobile home is perhaps closest to a completely industrialized unit because it requires relatively little site labor.

The pressing housing needs identified in developing nations and the continuing urbanization of the world guarantee that continued industrialization of housing will be important in decades to come. The United Nations has recognized the need for rural housing throughout the world, but especially in those countries that have relatively low per capita incomes in rural areas and that lack the technical knowledge for the construction of accommodations compatible with varying standards of living and climatic conditions.

The three main criteria of all housing requirements are population increase, replacement of existing stock, and the alleviation of the present shortage or deficit.

Because the expenditures for housing are so great, even small percentage savings represent substantial sums. For example, Nunnally (1973) notes that a savings of 1% in construction costs would represent a savings of over $30 million to the people of the state of Florida. During periods of economic recession and fiscal austerity, there will be increased need to examine increases in productivity that can be realized through industrialization of housing production.

The Degrees of Industrialization

Industrialization need not take place exclusively at a factory or site remote from the location where housing is to be constructed. Industrialization can be, and is, practiced at the construction site by homebuilders whose annual volume of work is small. In fact, the majority of the member builders of the National Association of Home Builders (NAHB) maintain relatively small operations. In a 1964 survey of its members, the NAHB noted that about 27% of its membership constructed 11 to 25 units, both single-family and multifamily, while 37% produced less than 10 units a year. These data indicate that industrialization, for the small builder, must be limited in many instances to onsite operations that are repetitive. In this sense, industrialization can be practiced at the construction site by small builders.

Naturally, the savings effected by industrialization vary widely. A study reached the general conclusion that the overall cost of prefabricated buildings is lower than that of conventional buildings by about 10 to 15% in countries that have amassed some experience with industrial methods of production and assembly of prefabricated components (Washington, 1969).

Often, smaller builders participate in industrialization by using prefabricated components. Prefabricated components used by builders include trusses, plumbing "trees," prehung doors, molded fiberglass tubs-and-enclosures, precast concrete wall and floor panels, and heat pumps. Use of precut studs and standardized sizes of lumber also permit small-scale industrialization to be used where custom designs are not necessary.

While many of the smaller builders advocate and use prefabricated components, local building code officials have not given complete support to their use and block their use in some locations.

Fragmentation of the housing market precludes, in some instances, adoption of more complete industrialization techniques. Economics of a scale that result from volume purchases, repetition of operations, and associated economy of construction are generally not available to small homebuilders.

At the other end of the industrialization spectrum are the widely accepted mobile and prefabricated homes. Prefabricated homes have received wide acceptance in U.S. housing subsidy programs, and mobile homes have grown substantially in size and sales since their widespread introduction following World War II. In 1967, mobile home production equaled 23% of all single-family, nonfarm housing starts (Washington, 1969).

Use of mobile homes and prefabricated homes obviously reduces the onsite labor normally associated with conventional "stick-built" housing. Restrictive zoning, building codes, and labor practices may not always be compatible with use of industrialized units. Usual claims for industrialized housing are that it increases the speed of construction, results in a better product made under supervised conditions, and allows a much greater output from the same labor force.

Where total capitalization is sparse, industrialization may be impossible to effect. The degrees of industrialization and the first-cost economies realized from them may be less important than savings that result from early occupancy of housing units and concommitant generation of rent income.

The median cost of housing in the United States in 1975 is approximately $38,000. Because the majority of homebuilders are nonunion firms, the hourly earnings for home construction labor are less than for comparable occupations in the building trades. A 1972 Department of Labor Bulletin (Washington, 1972) indicated that the most frequently quoted obstacles to efficiency were building codes, lack of skilled workers, and adverse work practices. Inclement weather was also identified as a persistent problem and a major deterrent to house construction.

The most frequently mentioned prefabricated items used in single-family houses were kitchen cabinets, vanities, preassembled windows, and prehung doors. Use of ready-mix concrete and gypsum products have reduced onsite labor requirements in housing construction and have shifted some jobs from construction into manufacturing. Man-hour requirements are declining considerably faster in manufacturing and other industries than in construction; however, according to the U.S. Department of Labor, the distribution of construction costs for single-family houses in 1969 was as follows: materials, 43.4%; overhead and profit, 29.7%; onsite wages, 20.4%; selling expenses, 2.7%; supplemental wage benefits, 2.7%; and equipment, 0.9%.

Prefabrication is also applied to multifamily housing. Construction of apartments, condominiums, and public housing has benefited by use of nearly complete prefabrication systems such as the Shelley, Panelfab, Echo, Balency, and Ucopan systems and others.

Industrialized housing production finds its greatest acceptance in areas where a large, aggregated market exists. Production of completely industrialized units has emanated from European and Asian nations where there has been a concentrated need for multifamily housing. In the late 1960s, it was reported that Eastern Europe concentrated on low-rise (3 to 5 stories) and high-rise (8 stories and above), multifamily structures (Washington, 1969). In Western Europe, the United Kingdom, Denmark, and Norway built over 50% of their housing as one- and two-family units while France, Sweden, and Finland concentrated on multifamily production. West Germany and the Netherlands split their housing production between these two types. In the United States, housing construction in 1968 was predominantly (61.7%) composed of one- and two-family units.

The industrialized systems used in Operation Breakthrough and in other commercial construction in the United States had, in many instances, origins in Europe. Capital-intensive industrialized systems require relatively large, sustained market demands to amortize the cost of equipment, which comprises a large portion of the first costs of industrialized housing.

The Federal Home Loan Bank Board released statistics (*Mobile Home Merchandiser,* 1975) showing that during October 1974, the average purchase of a new home required a $12,021 down payment and a $250 mortgage for 27 years. The average purchase price in October (1974) was $42,300.

Mobile homes constitute a large percentage of the lowest-cost new housing in the United States. Mobile home sales constitute more than 90% of the sales of new homes costing less than $10,000. The homes are completely industrialized except for site and foundation work and anchor installation. Double-wide mobile homes, normally 24 ft wide and approximately 60 ft long, are completely industrialized except for weatherproofing the joint and completing the foundation.

Semi-industrialization possesses a great deal of flexibility for the builder and developer than may not be met by full industrialization. It is difficult for any entrepeneur with a long-term view to consider entering the housing field on a large scale in the face of the variation in the amount of funds available for financing these programs from one year to the next (Washington, 1969).

While industrialization can benefit housing construction by reducing costs, it is generally recognized that other factors have a strong influence on housing costs. Land costs, especially in urbanizing areas, have escalated dramatically. Land speculation has added substantial costs to the price of a home. It has been suggested that a land reserve be established for low- and middle-income housing inside and outside of urban areas.

The percentage of onsite labor has diminished over the past decade. The effect of increased interest rates on housing costs has been an incentive to reduce construction and onsite labor costs through industrialization; however, small increases in interest rates accumulated to significant

TABLE 4.1. Effect of Interest Rate on Cost of a $20,000 Loan Over a 25-Year Period

Interest Rate (%)	Monthly Payment (Principal Interest	Total Interest (Over 25 Years)
6	$129	$18,600
6½	135	20,440
7	141	22,390
7½	148	24,330
8	154	26,280
8½	161	28,200
9	168	30,220
9½	175	32,370
10	182	34,460

amounts over the life of the mortgage. Table 4.1 shows the affect of interest rate on the cost of a typical, modest loan for 25 years. Naturally, the cost of adequate housing is itself a subjective term. Costs for housing would be higher in Tokyo or New York than they would be in rural areas of the southern United States.

As income rises, the ratio of shelter to consumer expenditures or to income declines sharply (Committee on Banking and Currency, 1971). Data in Table 4.2 illustrate this fact.

The extent of prefabrication of housing can vary widely as noted previously. The construction of single-family houses is a major component of national business and creates jobs at the construction site as well as in the

TABLE 4.2. Ratio of Rent to Income, by Income Class, 1965 and 1969

Income	1969	1965
$ 3,000–4,999	18	18
$ 5,000–7,499	15	15
$ 7,500–9,999	12	12
$10,000–14,999	11	10
$15,000 or more	<11	<10

mining, trade, transportation, and service industries that interact with the housing industry. In rural areas, where housing is not dense, local builders will construct only a few units each year. Homebuilding construction is varied and relatively unmechanized in rural areas. The labor force used by homebuilders has a high percentage of manual craftsmen and laborers.

Homes built in metropolitan areas require fewer man-hours per $1000 of construction cost than those built in nonmetropolitan areas. On a 100-ft^2 basis, it was noted (Department of Labor, 1972) that 71 man-hours were required in metropolitan areas. Table 4.3 shows that distribution of onsite man-hours for new private single-family houses by type of contractor in 1969.

Naturally, those contractors and trades constituting the largest percentage of construction labor costs are most likely to be reduced by industrialization or semi-industrialization. An example can be made using carpenters. Driving nails by hand takes considerably more time than driving nails by use of either an explosive-aided hammer or a pneumatic hammer. While carpenters have not been eliminated they have received new tools that make them more productive. Semi-industrialization also can take

TABLE 4.3. Distribution of Onsite Man-Hours for New Private Single-Family Houses, 1969

Type of Contractor	Percent of Onsite Man-Hours
General	31.3
Carpentry	16.9
Plumbing, heating, ventilating, and air conditioning	8.7
Painting and paperhanging	6.5
Masonry and stonework	9.0
Concrete and stucco work	7.2
Electrical (except heating)	3.7
Plastering and lathing	1.4
Roofing and siding	2.0
Ceramic tile, terrazzo, and marble work	2.2
Excavation and grading	1.6
Wood flooring	.9
Other flooring	1.6
All other types	6.9
Wallboard (included in all other)	(4.0)

TABLE 4.4. Average Hourly Earnings and Average Hourly Union Wage Rates Paid for New Private Single-Family House, 1969 (Department of Labor, 1972)

Occupation	Average Hourly Earnings	Average Union Hourly Wage[a]
Carpenter	$4.11	$5.35
Painter	3.99	5.01
Bricklayer	4.76	5.63
Plumber	4.85	5.73
Cement finisher	4.53	5.12
Electrician	4.86	5.57
Plasterer	5.02	5.34
Sheet-metal worker	4.60	5.48
Roofer	4.78	5.11
Operating engineer	4.36	NA
Tile setter	4.14	5.25
Soft floor layer	4.68	NA
Laborer	2.93	4.05
Helper and tender	2.56	4.05
Truckdriver	3.54	NA
All occupations	3.94	5.14

[a] NA—not available.

advantage of the use of semiskilled or unskilled labor rates in place of skilled labor rates. In some instances, it is possible to substitute nonunion workers for union workers and thereby reduce costs associated with higher union wages (Table 4.4).

Capital improvements in land values also may be considered as one step in promoting semi-industrialized housing. For example, utilities (except sewers), streets, sewer hookup, curbs and gutters, sidewalks, community recreation facilities, storm drains, and street lighting are often included in with subdivided or developed land. Housing location can be controlled and unit costs decreased with prior inclusion of these improvements. The U.S. Agency for International Development has assisted housing programs in several developing nations by providing site and services facilities.

The U.S. Department of Labor noted (1972) that several factory prefabricated items have been widely included in houses. Table 4.5 includes a list of these items that are now widely used and contribute to savings and may be considered semi-industrialization.

TABLE 4.5. Prefabricated Items Included in Surveyed Houses, 1969

Item	Percentage of Houses Containing Item
Preassembled windows	78.4
Kitchen cabinets	77.6
Prehung doors	64.4
Bathroom or bedroom dressing vanities	46.4
Precut lumber	30.4
Offsite fabricated ductwork	28.4
Staircase units	24.8

Use of prefabricated plumbing and glass fiber reinforced polyesters for molded bathtubs with integral wall is growing. In fact, some homes and apartments are constructed by conventional techniques except for inclusion of completely prefabricated bathroom or kitchen units.

Dietz (1974) summarizes one method by which industrialized or manufactured housing can be described. Semi-industrialization may involve "precutting" of housing components to reduce onsite labor and erection time. Another type of semi-industrialization process involves "frame and infill" techniques, which permit smaller prefabricated panels to be installed in a structural frame. In another technique, the precutting of all elements and partial assembly into shell elements results in panelized houses.

Three-dimensional volume elements or large boxes or modules are sometimes used. A minimum of onsite labor is required to complete the construction although specialized equipment may be required to erect heavy modules.

Mobile homes are the ultimate in single-family industrialized housing. Within the past 10 to 12 years, the production of mobile homes has increased approximately 300%. The economy effected through large-scale purchase immunity from delays caused by weather, and high degrees of productivity and organization have enabled prices for mobile homes to remain approximately half of those of traditional site-built housing.

The movement toward industrialized housing is composed of many separate trends and this obviously provides substantial flexibility to homebuilders who wish to incorporate a product or process of industrialization. Especially where the demand density is low, as in rural areas, the builder wishes to be free of capital-intensive activities that require more sales for amortization than the market area can reasonably be expected to produce. Glazed tile, precut lumber, prehung doors and windows, and use of standardized components can be expected to assist in meeting the world's serious housing shortage.

Prefab Homes at Market

Prefabricated homes include those units that have significant portions of their structure precut and/or preassembled for rapid erection on site. The development of prefabricated homes has been rapid and generally responsive to the findings that more and better housing is needed at lower cost.

The direct labor content of housing can be reduced significantly through industrialization or prefabrication. Robert Platts (Dietz, 1971) has observed that the direct-labor content of wooden-frame project housing constructed in the northern United States and Canada is approximately 25% of the total product cost. For this same type of construction, materials comprise approximately 57% of the cost and overhead and profit amounts to 18%.

C. F. Dally has estimed (Reidelbach, 1970) economies that result through increased productivity of modular units. To establish a modular facility requires, in addition to extensive capital, attention to the following 15 items identified by Dally:

BASIC PROCEDURE TO ESTABLISH MODULAR FACILITY

1. Market study, complete and detailed.
2. Plant location and site established.
3. Engineering, systems and methods.
4. Plant equipment ordered and scheduled.
5. Interviewing and hiring—administrative, clerical, indirect and direct personnel.
6. Setting up office space and equipment.
7. Sales programming and advertising; public relations program established.
8. Equipment installation, plant training given to a nucleus of crew.
9. Assembly lines engineered, set up, and readied for production.
10. Additional plant personnel hired; scheduling of materials.
11. Standard unit in trial production run.
12. Final training completed; time quotas established.
13. Materials arriving; subassemblies in progress; inspection and time studies evaluated.

14. Line production begun at ¼ rate (30 to 45 days elapsed since item 5).
15. General production under way; future production projected and scheduled with materials suppliers.

Estimated percentages of total cost for material, profit, labor, and other cost categories are shown for various production rates in Table 4.6. Even in prefabrication, economies of large-scale production are evident.

The stresses that occur in plant handling, in transportation, and erection of prefabricated units may require special attention. Modular units must be constructed so that the units can be joined efficiently at the job site. One constraint, for example, on a modular or prefabricated manufacturer is that the highest point of a prefabricated unit not exceed 13 ft 6 in. Consequently, flat-roof pitches are commonplace in modular units.

The total number of prefabricated homes available at the market has increased significantly in recent years. The component of this type housing that has experienced the greatest growth is the mobile home. Table 4.7 shows the total number of units produced in the United States from 1955 to 1971.

Approximately 5% of mobile home production is for purposes other than residential. Some units are used for offices, schools, banks, and other business and commercial uses. In 1970, reportedly 5% of the new mobile homes produced were bought as second homes.

Mobile homes have grown in market acceptance and in size. The 12 ft wide unit and double-wide units are very popular and are widely manufactured in 60 to 65 ft lengths. Drury (1972) has noted that, in 1955, the average size of a mobile home was 8 by 45 ft or 360 ft^2. In 1966, the average size was 12 by 60 ft or 720 ft^2. This represents a 100% increase in livable

TABLE 4.6. Estimated Percentage of Total Cost

Cost Categories	1 House/Day	2 Houses/Day	4 Houses/Day
Material	46.0	46.0	46.0
Profit	15.0	19.5	23.5
Labor	16.5	15.0	14.0
Administration	9.5	8.0	6.5
Sales	7.5	7.0	6.5
Manufacturing expenses	5.5	4.5	3.5
	100.0	100.0	100.0

TABLE 4.7. Production of Mobile Homes, 1955–1971

Year	Total Units Produced
1955	111,900
1956	124,330
1957	119,300
1958	102,000
1959	120,500
1960	103,700
1961	90,200
1962	118,000
1963	150,840
1964	191,320
1965	216,470
1966	217,300
1967	240,360
1968	317,950
1969	412,700
1970	401,200
1971	485,000

Source: National Manufactured Housing Federation, Washington, D.C.

space compared with an increase in price of approximately 31%. In 1970, the average cost of a mobile home was approximately $8.35/ft^2. Table 4.8 lists the average purchase price of mobile homes and conventional single-family homes from 1960 to 1970.

The growth and marketing of wider mobile homes have paralleled the growth of highways, especially the interstate system. Table 4.9 shows the sizes of mobile homes produced, by percentage, for each year between 1960 and 1970. When mobile or modular homes are transported by truck within a 300-mile radius of the manufacturing plant, moving costs will generally account for less than 3 to 5% of the finished cost of the dwelling, excluding the price of the lot. The average maximum shipping radius and the average maximum shipping radius in 1969 (Reidelbach, 1970) 223 miles and 534 miles, respectively.

TABLE 4.8. Average Purchase Price of Mobile Homes and Single-Family Houses

Year	Average Purchase Price of	
	Mobile Home[a]	Conventional Home[b]
1960	$5000	$13,800
1961	5600	13,875
1962	5600	14,325
1963	5600	14,875
1964	5600	15,575
1965	5600	16,150
1966	5700	16,750
1967	5700	17,325
1968	6000	18,525
1969	6050	19,225
1970	6110	18,325

[a] National Manufactured Housing Federation, Washington, D.C.
[b] U.S. Department of Labor.

The cost of mobile homes is, on the average, less than $10,000 and consequently affordable by a larger segment of the population. Often, young families find the mobile home an attractive alternative to apartment living. The mobile home is affordable, and at a cost of approximately one-third or less than that of conventional housing, permits the young family to sample

TABLE 4.9. Sizes (%) of Mobile Homes Produced, 1960–1970

Width (ft)	Year										
	1960	1961	1962	1963	1964	1965	1966	1967	1968	1969	1970
8	9.5	1.9	2.0	1.5	0.9	0.9	0.5	0.3	0.1	0.5	0.3
10	90.5	98.1	72.7	73.3	59.8	41.4	24.6	7.3	2.2	1.1	0.5
12	—	—	5.0	6.4	21.1	45.1	65.3	84.1	85.9	84.2	8.1
14	—	—	—	—	—	—	—	—	—	2.3	8.1
16 and double and triple	—	—	20.3	18.8	18.2	12.6	9.6	8.3	11.8	11.9	12.5

Source: National Manufactured Housing Association, Washington, D.C.

homeownership and have equity in an investment. The mobile home has been attractive to young couples who find that their finances cannot pay for their tastes. Older people, including the retired, find the small space and efficiency of mobile homes to be compatible with reduced family size and disposable income.

The number of mobile home, modular, and prefabricated home manufacturers has declined in recent years as a result of the elimination of marginal producers. This has been caused by poor management, competition, and transportation costs, together with the relatively recent trend toward the acquisition of small companies by larger firms as noted by Drury (1972). It is important to note that the cost of the average mobile home includes major appliances and furnishing. Total costs of low- and medium-priced conventional homes are still approximately 45 to 65% higher than the per square foot costs of the average mobile home.

Advantages of Semi-Industrialization

In general, industrialization implies making the best use of readily available resources, achieving the longest possible uninterrupted production runs, and seeking repetition of effort and greater efficiency and productivity. Industrialization makes use of skills built into machines. Process workers are more easily trained for repetitive tasks than are tradesmen who cannot be multiplied to meet peak demands. More standardized components and more homogeneous quality result from semi-industrialization.

Development of industry, trade, and tourism in rural areas could alter the ideas of the rural population and the need for housing and repaired housing will grow. Semi-industrialized housing is advantageous because it can be adapted to rehabilitation as well as to new construction. Replacement of windows and doors or plumbing components is enhanced when standard dimensions and products can be used. A fundamental case for building by means of component units derives from the need to reduce work on the building site where material delivery delays and inclement weather slow progress. Semi-industrialization is warranted where sufficient demand for a product or component exists.

Industrialization and semi-industrialization of housing endeavors to save the labor force required. The use of semi-industrialization in housing production takes advantage of the learning behavior and economies that result from repetitive performance of tasks. Let us examine one method for characterizing the time required to manufacture a house component, such as a window and its frame. Figure 4.1 shows how the time per unit can be expected to diminish as the number of units manufactured increases.

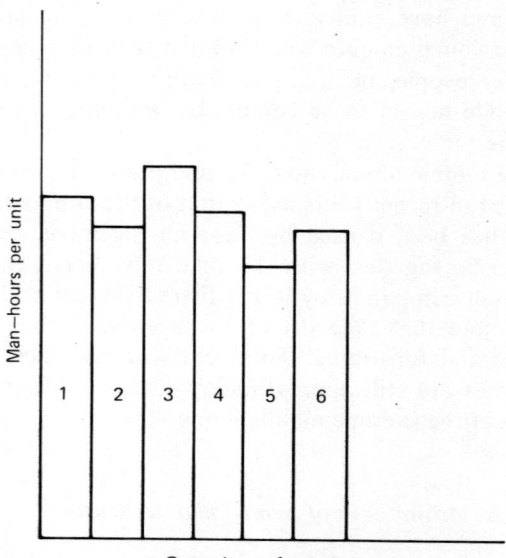
FIGURE 4.1. Unit production time related to number of units produced.

The total time correction, T, may be expressed as

$$\lim_{B \to \infty} \Delta T = L \cdot e^{-B_0/L}(1 - e^{-B/L}) = Le^{-B_0/L}$$

The speed of the process at any time (B) may be expressed as

$$\frac{dT}{dB} = 1 + e^{-B_0/L} e^{-B/L}$$

Hence the intial speed is

$$\frac{dT}{dB} = 1 + e^{-B_0/L}$$

The speed of the basic process is always 1, and the expression $K = e^{-B_0/L} \cdot e^{-B/L}$ may be taken as the speed of correction. Figure 4.2 depicts the appropriate geometry.

In general, the time required to complete a component can be represented by the following formula suggested by Hugsted (*Towards Industrialized Housing*, 1966):

$$T = B + Le^{-B_0/L}(1 - e^{-B/L})$$

where T = total time, including correction
  B = basic time
  B_0, L = parameters determining the initial speed and time correction to be made

The marked feature of many of the developing countries of Africa, Asia, and Latin America is their rapid rate of population growth. In these developing continents the annual average housing requirements to meet population increases, present housing deficits, and housing obsolescence over the next decade was estimated at 8 to 10 dwelling per 1000 inhabitants. This rate was only achieved in a few of the more advanced regions of the world. Only by use of semi-industrialized and prefabricated housing can the world need for housing be seriously addressed.

The United Nations defined *industrialized building* as "a continuity of production implying a steady flow of demand; standardization; integration of the different stages of the whole production process; a high degree of organization of work; mechanization to replace manual labor wherever possible; research and experimentation integrated with production." It defined *prefabricated building* as "the transfer of varying propositions of the operations of manufacture from the building site to factories or workshops, which may be independent of the site or associated with it. In this connection the term 'partial prefabrication' is sometimes used."

The cost of new houses in the United States has increased approximately 16% from 1970 to 1974. The average sales price of a new home in 1978 was $59,500. In 1974 it was $41,300. Advantages of industrialization include reduction of costs that have been noted to be high in developed nations and

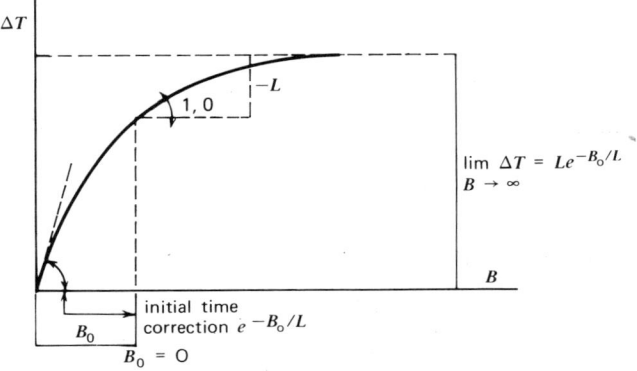

FIGURE 4.2. Geometric interpretation of the learning equation.

138   Semi-Industrialized Housing Production

constrain just as seriously the provision of adequate housing in developed countries.

Semi-industrialization may be practiced widely in rural areas and for small markets where demand for housing could not support the capital-intensive activities associated with full industrialization.

More rapid onsite erection and economies that result from reduced interest and overhead expenses can be considered a major advantage of semi-industrialization. Because semi-industrialization does not represent, in many cases, a serious discontinuity with traditional building practice it is acceptable by workers who might otherwise be hostile to major changes in construction practice. Where semi-industrialization enables the builders to enclose their structures more rapidly than without industrialization, it provides more continuous work without interruption by inclement weather.

Increased quality control should be possible through semi-industrialization. For expedient shelter, such as might be required to replace housing destroyed by earthquake, typhoon, or other natural disaster, the semi-industrialized or fully industrialized house may be one of the only ways that housing needs can be met. Consumer familiarity with commonly used components can enhance housing.

Industrialized units have been present long enough to have received substantial public acceptance. Nevertheless, zoning ordinances and the scarcity of inexpensive land in urban areas make it difficult to consider these units for low-density developments in the central city. The adoption of semi-industrialized housing is not likely to be without problems.

5

Industrialized Housing Production

Edgar Wood

Industrialized housing production, as the title implies, is a catch-all phrase encompassing all types of structures manufactured offsite that range from mobile homes to multistory, high-rise apartment buildings. The materials used in these structures may vary from wood, steel, or concrete to combinations of materials each being used as the main element of a housing scheme.

Definitions

The discussion of industrialized housing production in the construction industry is resulting in a new vocabulary. As an aid to a better understanding of this new industry, the following defines some of the more common terms in use today:

Box System	May be made of two-dimensional elements assembled into a three-dimensional box or module at the factory. It can also be cast monolithically on special forms. These boxes are designed to function as structural elements.
Closed system	Is a proprietary system. It involves a negotiated contract because the package and price must be considered together.
Components	Are interchangable elements designed for open systems allowing the designer a freedom of choice without sacrificing unity.

Elements	Used generally to describe the concrete units manufactured in a factory for building systems.
Frame system	Is designed using standardized sections for the columns, beams, and purlins to a predetermined grid to form the structural portions of a building.
Industrialized construction	Describes systems building manufactured offsite in a factory using industrial process. It is also used interchangeably with "prefabrication" and "systems building." It is preferred over "prefabrication" as giving a better connotation for this emerging industry.
Mobiles	Refers to a lightweight box system manufactured in a factory complete in all detail and usually transported to a site on its own wheels. This unit is normally purchased complete with the furniture installed.
Modules	Is used in conjunction with subsystems used in an open system of construction. It can be used with three-dimensional structural subsystem modules or kitchen, bathroom, or mechanical modules manufactured in a factory.
Open system	Can use components or modules interchangeably with other open systems.
Panel system	Describes the majority of the European concrete systems, which are normally full-sized bearing walls that support a room-sized concrete ceiling panel. These concrete elements are all site assembled.
Post tension	Refers to the onsite tensioning of the special reinforcing used to bind precast concrete panels together and transfer stresses to meet design criteria.
Prefabrication	Used commonly when any standard-sized units are manufactured in a factory and rapidly assembled.
Prestress	Refers to introducing a stress into a special reinforcing prior to introducing the concrete. It is universally used in concrete manufacturing facilities that make floor panels, piles, and bridge sections.

Sectional modular	Is manufactured of wood in longitudual modulars. It is a living unit, complete in all detail, designed to keep within the allowable size limits established for units to be transported on the highways.
Subsystems	Are the structural, electrical, and mechanical components or modules designed to interface to form a complete system.
System	Describes the method employed for construction that uses a controlling set of disciplines that govern when a housing scheme will be planned, designed, scheduled, manufactured, and erected onsite.
Systems building	Is the application of modern industrial concepts of business and building technology, including computerization, to the construction industry, using a closely knit organization that includes architect, contractor, manufacturer, and skilled and unskilled worker.

The History of Industrialized Construction Immediately after World War II

The housing resources in many of the countries in Europe were in a desperate condition as a result of World War II. Cities like Rotterdam in the Netherlands were completely devastated, requiring almost a total commitment of their labor resources for reconstruction. This great need brought about the development of many successful housing systems in Europe. The lack of labor removed the conventional methods of construction so that the cost of competitive methods used in housing production was not a factor for consideration.

In this same period of time, the late 1940s, the United States was not experiencing the same pressures on its housing or labor resources; therefore, no concentrated effort was expended to develop this type housing production. One housing scheme worth noting, however, was developed at this time using federal funds supplied by the Reconstruction Finance Administration to build a single-story detached house of porcelain steel panels. This housing unit manufactured by the Lustron company was acknowledged as a quality unit with excellent features. Within a two-year period this operation was declared bankrupt.

Therefore, it is evident the United States and Europe did not experience the same growth in industrialized housing production in the postwar years.

During the period following World War II, the late 1940s through the 1960s, industrialization of housing units in the United States was primarily all single-family detached houses and mobile homes, with relatively no penetration into the European-type high-density living units manufactured of concrete. Europe on the other hand continued to produce housing with a good percentage of the housing units produced by using industrialized construction technology. According to data from the Department of Housing and Urban Development (HUD), in 1970 the following countries attained these percentages.

1. Bulgaria, 33%.
2. Czechoslovakia, 69%.
3. German Democratic Republic, 90%.
4. Hungary, 69%.
5. U.S.S.R., 57%.
6. Denmark, 35%.
7. France, 21%.
8. The Netherlands, 30% (1968 figures).
9. Norway, 28%.
10. United Kingdom, 38%.

It was not till 1969 that the U.S. government through HUD created Operation Breakthrough to bring about a change and improvement in the whole process of housing. The government invited private industry to submit proposals to develop a building system to produce housing to meet criteria developed by HUD. From 236 submissions, 21 national concerns and consortiums were selected. Money was provided by Congress to enable each of the winners selected to build their prototype units on 9 different sites in various parts of the United States for further evaluation. Valuable information was developed through this approach, but because of the lack of an aggregated market or need and continued governmental support, it is known that from among the winners of the HUD competition the number of companies engaged in industrialized construction today has been sharply reduced.

It is necessary to define the type of construction each of these companies used in their housing schemes to enable a comparison to be made of the industrialized systems of Europe and the United States.

First, it is interesting to note that of the final 21 companies in the group of systems selected by HUD, 15 were using Type VI construction, limiting them in height and area according to the model codes that regulate construction in the United States (Table 5.1).

The History of Industrialized Construction Immediately after World War II

Company	Type of Construction	Height Limitation
Alcoa Construction System, Inc.	VI	3 stories
Boise Cascade	VI	3 stories
Building Systems International, Inc. (Balency)	I	No limit
Camci, Inc. (Tracoba)	I	No limit
Christiana Western Structure	VI	3 stories
Descon/Concordia	I	No limit
F.C.E.-Dillion, Inc.	I	No limit
General Electric	VI	3 stories
Hercoform Marketing, Inc.	VI	3 stories
Home Building Corporations	VI	3 stories
Levitt Building Systems, Inc.	VI	3 stories
Material Systems Corp.	VI	3 stories
National Homes Corp.	VI	3 stories
Pantek Corp.	VI	3 stories
Pentrom, Inc.	VI	3 stories
Republic Steel Corp.	VI	3 stories
Rouse-Wates, Inc.	I	No limit
Inland-Scholz, Inc.	VI	3 stories
Shelly System	I	No limit
Townland System	VI	3 stories
TRW System Group	VI	3 stories

Five of the remaining companies allied themselves with foreign system builders using the big panel and box-type concrete systems that were adapted for this competition. Only one company developed a concrete system capable of meeting the most restricted requirements of model building codes. It must also be noted that not all the systems-oriented housing companies that existed in the United States at that time submitted a proposal to HUD. There are total building systems developed in this country using Type I construction that are based on industrialized building technology, which, coupled with their proprietary programs, make them among the most sophisticated of their kind in the world.

The second condition to examine is the labor resources of any country and how it relates to industrialized construction. In the previous section it was shown that generally in Europe a need for housing was developed as a result of the tremendous depletion in their housing resources. This resulted

TABLE 5.1. Group A Residential Occupancy—Height and Area Restrictions

Type of Construction Used	Allowable Heights — Story Height	Allowable Areas (ft²/floor) — First Floor	Second Floor	Third Floor and Above
Type I—fireproof	No limit	No limit	No limit	No limit
Type II—fire resistive	80 ft	No limit	No limit	No limit
Type III—heavy timber	3	18,000	15,000	9750
Type IV—noncombustible	5[a]	12,000	10,000	6500
Type V—ordinary	5[a]	12,000	10,000	400
Type VI—wood frame	3	7,000	7,000	6400[b]

[a] When over 4 stories in height, two-hour fire-resistive floors shall be required over basement or cellar.

[b] One-hour fire-resistive construction shall be required throughout.

in the overtaxing of their labor resources. Therefore, industrialization was the only solution available to solve their problem. Upon closer examination of the labor conditions in Europe it is found that countries such as the Federal Republic of Germany, with its high economy, had to import its labor force from other countries, including Spain, Italy, and Turkey. This labor force is estimated to represent 11% of their total population. During 1964, in Switzerland 31.6% of the construction labor force was imported. Conditions such as these, coupled with a reduced population growth, require the aid of new technological developments toward achieving housing goals as shown in Table 5.1.

The portion of our labor resources that are organized have reservations about the use of industrialized construction for housing production because it means a change must take place if they are to participate. To fully accept industrialization it means work rules and traditional trade practices must be changed to meet the disciplines governing industrialized construction. A prime example would be the experience of a large systems company in Scotland that had to locate a construction trade union which would recognize a new trade designation, Panelmaker. In their manufacturing process, the duties of the panelmaker include cleaning the steel molds, placing the prefabricated steel mesh, windows, door frames, and electric conduit in the molds prior to his or her placing the concrete and giving the exposed surface any special treatment required. Traditionally, this would require workers representing a minimum of five different trades. The key element involved in this case is that all the material placed by this worker was prefabricated, requiring no additional work from him or her other than the placement of the concrete in the mold. This is a change in traditional work practices—the giving up of work that was reserved for a specific trade.

Reservations of the members of the building trades that industrialized construction will reduce their job opportunities are groundless for many reasons, among them, (*a*) every country of the world has increasing large housing deficits; (*b*) studies conducted by two major U.S. firms noted relatively limited net displacement of skilled labor because of industrialized construction; and (*c*) studies in England showed the proportion of carpenters and joiners increased, masonry work declined, and plumbing remained the same.

The third condition to examine is the building controls, a determining factor in the selection of building systems for a construction project. Table 5.1 shows the six types of construction used for residential occupancy in the United States. Normally, only three of these designations are used for residential construction: Type I—fireproof, Type II—fire resistive, and Type VI—wood frame. The table shows no limit in the story height or floor area for Type I, while Type II is limited to 80 ft in height, but no limit on the floor area. For Type VI the story height is limited to 3 stories and the floor area at 7000 ft^2/floor.

Type I construction is normally all reinforced concrete or a steel frame encased in concrete with concrete slab floors. The exterior walls have to be designed to meet the most rigid fire ratings. Therefore, they are normally of concrete if it is a building system or brick masonry if constructed using conventional methods. This type of construction qualifies for being built in the most restricted fire zones of large cities and buildings of any height.

Type II construction is a steel-framed building with its columns and beams fireproofed by encasing them in concrete or other approved material. The floor system may be framed open-web steel trusses or joists or some other approved system that is then only protected on the underside by the ceiling material. This type construction is therefore limited in its story height, averaging 8 stories.

Type VI construction being a wood frame and combustible has many limitations. It is limited in height to 3 stories and relatively small floor areas. This type construction does not qualify for projects in core cities or in the more restricted fire zones.

The last condition to examine is cost. Starting with any systems that are qualified at Type VI, one can readily envision that any system in this category would have to be extremely efficient to compete in cost with any small contractor whose overhead is nil as compared to maintaining the overhead of a factory and equipment costing in excess of $1 million coupled with inventory, transportation, and in many cases the additional cost involved for an onsite contractor. The more successful systems in this category would have to be the mobile home manufacturers, although the product is visibly inferior in quality to conventional construction.

The remaining Type I and II construction can be examined together if a series of assumptions are made. One is that Type II construction, because of the requirement of encasing the steel frame in concrete or other fireproofing material, can best be erected conventionally. Therefore, it may be assumed that this type building is the competition of building systems for up to 8 stories in height. Some of the labor required onsite for this type construction are highly intensive in the following top pay-scale categories: (1) ironworkers, (2) masons, (3) carpenters, (4) plumbers, and (5) electricians. The last assumption being made is that for the purpose of this comparison the building being examined is serviced by one crane. This fact allows a comparison to be made on the time required to hoist all materials used during construction, which can be equated to productivity and finally cost.

Type I systems, using concrete modules and elements for its structure and three-dimensional interior modules containing the kitchen, bath, elevator, and mechanical subsystems—all being manufactured in a controlled environment—enjoy obvious advantages over the Type II conventionally built building. For ease in comprehending how these result in economic advantages, some of the more apparent ones are listed.

1. Purchasing practices enable systems manufactueres to get much of their material direct from the material manufacturers as an Original Equipment Manufacturer account. This is the lowest price structure available in the industry.
2. Labor rates of the employees in the factories are less than onsite labor rates.
3. The number of workers required onsite is reduced (see Table 5.2).
4. Productivity is much higher in manufacturing facilities.
5. Construction time is reduced as shown in Table 5.2.
6. One lift of the crane to position a three-dimensional concrete module weighing approximately 25 tons is equal to 40 lifts of the crane to accomplish the same amount of work on a conventional job.
7. Kitchens, bathrooms, and mechanical modules are manufactured complete in all detail, including all fixtures and appliances ready to connect onsite into a vertical utility case.

As noted in Table 5.2 many European countries have a history of savings in labor force and elasped construction time when various industrialized construction systems are used, resulting in the reduction of construction costs.

Types of Industrialized Construction

The many types of industrialized building programs used today to provide housing are too numerous to fully describe in this chapter. Many large housing developers have developed their own "system" for the particular type of housing scheme they market. Many of their components are manufactured in an onsite facility where the subcontractors perform their work. These are normally made in full story-high units and installed on the foundation with a crane. One well-known developer manufactures 10 housing units of this type per day or 2500 per year equaling or exceeding the production of many well-known building systems available today. Operations of this type are described as "vertically postured," where the company controls the site acquisition, land planning, manufacturing, construction and provides its own marketing services.

The "horizontally postured" companies in the housing business are those that manufacture components for the structure and the interior subsystems and market these products to others who use open systems for their housing scheme. Included in this group may be the mobile home, panelized, modulars, box systems, or others when they are marketed in this manner.

The various types of industrialized building systems when grouped generically would result in three groups: the frame, panel, and box.

Examining each of these three groups to determine their composition, it is found that the frame system would include many of the prefabricated steel buildings used for storage and manufacturing facilities by industry and such systems as the School Construction System Development which originated in California.

The panel system would include many of the wood precut and prefabricated units sold in the United States by manufacturers and lumber dealers. These would include such components as roof trusses, panel walls, and floor systems. Some steel companies have proprietary framing systems that fall into this classification. European-type concrete systems whose components are produced and delivered as two-dimensional elements are in this category.

The box system would include wood modulars, mobile homes, and any concrete system that manufactures and ships its components in a three-dimensional shape.

Within this last group is the mobile home, and it must be examined to learn the impact that this type of industrialized construction is making on the housing market in the United States. For 1978, it is reported that

TABLE 5.2. Percentage Reduction in Onsite Operation and in Total Building Costs in Industrialized Building Systems—As Compared to Conventional Building Systems in Selected Countries

Country	Onsite Completion Time	Onsite Man-Hours	Total Onsite and Offsite Man-Hours	Total Building Cost
1. Bulgaria	50	—	35	—
2. Czechoslovakia	75	50	30	15
3. Denmark	—	47	25	15
4. Finland	2–3 months earlier	33	—	5–10
5. France	—	50	—	10
				3
6. Federal Republic of Germany	—	50–55 (Hamburg)	—	5—up to 9 stories (Hamburg)
				12—above 9 stories (Hamburg)
				10
7. Netherlands	—	—	No reduction	—

8. Norway	—	—	No reduction
9. Poland	53	—	10–17
10. Romania	—	30	—
11. Sweden	—	—	No saving
12. U.S.S.R.			
a. large panel	50	35–40	14–15
			8–10; 15–20
b. box type	75–85	15% below panel system	Same as panel system
13. United Kingdom (large panel)	58 cf. to small site 35 cf. to large site No savings—cf. to rationalized conventional	—	Cheaper—above 6 stories Same—3 to 6 stories More expensive—1 to 3 stories

Source: U.S. Government Printing Office, 1972, 0-472-949, Springfield, Virginia.

275818 mobile homes were manufactured, accounting for 25% of all single-family housing units produced in the United States that year. The industry reports a trend toward units up to 80 ft in length and 24 ft in width. These units, because of their width, are called "double wides." They are transported as two units and joined together at the building site. For 1978, 30% of all mobile homes sold in the United States were double wides.

To familiarize the reader with the different type buildings constructed using concrete, let us consider the Thamesmead project in England. The industrialized construction techniques used in this project were developed by the Balency system. The project was designed to house 60,000 people as part of a relocation program in conjunction with the urban renewal projects in London. After being designed by architects, this project was then awarded to a systems company for construction and therefore did not contain the stereotype appearance resulting in most Euopean systems. The Balency company erected a special facility adjacent to the construction site. This industrialized concept is classified as a panel system and had the majority of its elements case on tilt tables, which were hydraulically operated and had a heated cover. Using this heating technique to accelerate the curing of the concrete element, the company was able to cast two elements per day on each mold.

Another well-known European company, Larsen-Nielson produces a readily identifiable panel system. The equipment used to produce their vertical cast and vertically erected elements are battery molds. These molds produce concrete elements that, in many cases, are smooth enough to receive paint or paper without any further treatment. The concrete floor elements are produced in a special mold that contains pipe mandrels which produce horizontal voids in the finished product. These voids reduce the dead weight of the concrete element by as much as 40%. It requires 15% insitu concrete for buildings of this type to top the floors and fill in all voids used under and around each panel to connect it onsite. This fact, coupled with the magnitude of bracing required, precludes the use of prefabricated three-dimensional interior subsystems such as kitchen and bathrooms. The concrete depositing machine not only fills the molds, but also contains vibrating roller screeds that finish the element to a true smooth surface. The vacuum lifter is used by the operator to stack this hot concrete (140°F) element on transport carts to a height of about 8 ft. This equipment is used to produce a concrete element every $5\frac{1}{2}$ min or about 90 panels per work shift. This clearly demonstrates the productivity of a capital-intensive facility.

The concrete elements used in the ECHO-Modular system can be manufactured using various equipment. It is then assembled into three-dimensional modules at the factory and transported to the jobsite. When erected, by a five-man crew, these three-dimensional modules become the structure of the building. Because of the stability of these three-dimensional concrete modules, a minimum of bracing is required, for safety purposes only, thereby allowing the use of prefabricated kitchen and bathrooms to be installed in sequence with the structural modules. Upon completion of the erection sequence, the onsite construction trades finish the building process in a short period of time.

The interior subsystems for this highly industrialized system are built under license by others.

Advantages and Disadvantages of Industrialized Construction

The construction industry is recognized as the most fragmented industry in the world. Because of many constraints such as archaic building controls, union labor practices, and the lack of governmental support and aggregated markets, the building industry has been very slow to develop new computer-controlled production-line methods or techniques that have accrued to the appliance and automotive industries. There are buildings still being erected of brick using the same techniques similar to those used for the past 400 years.

As the United States and other highly industrialized countries try to control their population growth, the labor market diminishes. If a country maintains a zero-population growth for one generation, two-thirds of its population will then be supported by approximately one-third. This fact, coupled with the need to maintain an expanding economy, will bring about a dramatic change in the construction industry. The present cost of residential construction in the United States is making home ownership a luxury beyond the means of much of our population.

Therefore, the only solution of producing much-needed housing in the volume required and for a price that will reach much of the population of any country is through new technology. This technology must enable us to attain a very high percentage of productivity in all phases of housing construction to enable a reduction in costs.

The mobile home industry clearly demonstrates an advantage in using industrialized construction. Many people do not recognize the mobile home as a permanent home or structure, but this industry has made dramatic inroads into our housing market. Ten million Americans now occupy mobile homes. A closer examination of how this industry has developed shows some interesting facts:

1. Considered a vehicle under the law, the mobile home does not have the constraints of conforming to the thousands of local building codes in the

United States. The mobile home industry has its own performance code (ANSI A119.9, which is now law in 44 states).
2. Since mobile homes are considered a vehicle on their own chassis and are not considered permanently affixed to the site, they enjoy special low taxation in the form of an excise tax.
3. Mobile homes are normally purchased fully furnished, with everything incorporated in one finance payment.
4. Because of a lack of local building code constraints, the manufacturers can produce a long run of modules, using assembly line techniques, to the same specifications, resulting in substantial cost savings.
5. The cost to the consumer for ample living space is sharply reduced from conventional construction.

The manufacturers of wood panelized homes have demonstrated the efficiency of producing their products in a volume market. With the aid of computer controlled machinery they can produce trusses for roof systems and market them for less money than what it costs contractors to build on the site. These same manufacturers have developed a substantial market in the small house contractor who buys their products and completes the onsite erection.

The disadvantages of industrialized construction lie in three areas: governmental, physical, and financial support. In the governmental area is the need for government to assert leadership and enact a single set of building controls to supercede the many local codes that are not designed to keep up with the new technology and materials being developed. With this type support, manufacturers may invest more in research and development for new materials and methods if they had but the one set of controls for guidance.

Coupled with this type support is the need for the governments of different countries to combine housing markets that will insure the success of establishing industrialized construction facilities. The housing market in many small countries with large housing deficits can only be combined by their government.

In the examination of the physical aspects of industrialized construction the factors that may be considered a disadvantage would be size and weight. When transporting housing components over the road systems there are size limitations that must be observed. These rules apply to all states, with some modifications. Generally, a height limit of 13.5 ft has been established. In width, any unit wider than 8 to 10 ft requires a special permit. Units from 10 ft in width to a maximum of 12 ft require a special permit and an escort vehicle with a sign and lights. Transporters are also limited to traveling in certain day light hours.

The problem of weight can be a disadvantage on any job when the weight of the elements involved exceed the ability of the onsite labor to move or position components using hand labor. At this point, the size of the elements may be increased to the full capabilities of the special lifting devices or cranes.

In the construction of Type I building schemes using concrete elements the weight of these elements can be considered a disadvantage if special cranes have to be obtained for hoisting. This advantage is magnified when the height of the building exceeds the practical height conventional cranes can lift. When a heavy concrete module is to be erected this limit may be around 22 stories.

The other disadvantage of industrialized construction to consider is financial. The liability involved in establishing a manufacturing facility and purchasing equipment deters many. To support all the indirect costs involved in the manufacturing process each company must produce between 30 and 40% of its ultimate production capacity to arrive at a break-even point and still remain competitive. As most manufacturing facilities are established based on some financing plan, which normally is 20 years for the real estate involved and 5 for equipment, this would limit the number of companies available having this type of financial responsibility. Along with this is the cost of marketing the finished product as discussed in this chapter.

In conclusion, it is obvious that a dramatic change in the manner of construction of all buildings has been taking place in the last decade. In this period of time, many building materials have been manufactured to standard sizes, which enhances the development of industrialized processes. Also, special mechanical equipment has been developed for onsite use to increase productivity of each building trade. The scope of building programs have also increased in this same period. Therefore, it is safe to assume that the next dramatic development will be the total production of housing units using industrialized construction technology.

6

Building Systems Software

Iraj Majzub

Since the advent of the Industrial Era in the middle of the nineteenth century, we have witnessed a tremendous technological advancement in all fields of human endeavor. In every major field, people have been successful in applying the industrialized process of production to achieve higher productivity, better quality, and lower cost. Mechanization has allowed us to produce more and better appliances, clothing, household goods, food, books, and vehicles.

We have been able to cope with the increasing demands and with the higher expectations in practically every field, with the exception of construction, where although the expectations and demands have increased enormously, our techniques and methods of production have not changed at a parallel pace to cope with our needs.

Today, in fact, we are faced with shortages in housing as never before in our history. UN statistics indicate that millions of families in the world have no shelter at all, and approximately half the world population lives in indecent, unsanitary, and inadequate conditions.

Projections for the future are even more bleak: most experts estimate that the number of homeless families will be in the order of one billion by the end of this century if present trends continue.

The Promise of Industrialization

If industrialization and mechanization have been successful in providing us with other goods at an ever-increasing rate and scale, why not apply it in housing? This question has been asked frequently during the past several decades, each time opening the field to more questions than answers.

It is understandable why so much emphasis has been placed on the industrialization of housing or "systems building."

To better comprehend systems building and what it implies, a brief outline will be given in this chapter; moreover, systems will be classified and the concepts of industrialization will be elaborated in the following chapter.

Systems building, as stated previously, is a relatively new term in the building industry, indicating the industrialized process by which components of a building are conceived, planned, fabricated, transported, and erected onsite. It is an interdisciplinary field of activity and does not entirely rely on the technological aspects of the process. Both the software and hardware are stressed in systems building and their balance combination is a prerequisite for the success of this process.

The Question of Cost

Economists would agree that the use of the term "cost," separate from "value," is quite insignificant. In the field of housing, the term "low cost" is not descriptive of any quality.

We must endeavor to lower the cost while preserving the quality. Only in this way can we achieve higher performance.

It is in this light that systems building should be measured and studied since its primary purpose is to increase productivity through the use of mechanization and industrialized processes or systematic approaches in planning the use of people and equipment, while controlling costs.

As housing represents the largest investment a family undertakes, we should be more concerned with the economic value of the home rather than its cost. Systems building in general does provide economic value. But in the area of cost, the costs of systems at times are higher or, at most, competitive to conventionally built buildings of the same type. This is a very important point.

The advantages of systems building are in general not in lower cost, but in the following:

1. Higher productivity.
2. Higher quality.
3. Year-round production.
4. Quick assembly (erection).

A study of systems in Europe shows that the reported effect on costs was very small and was achieved mainly in high-rise construction. (See Table 6.1).

TABLE 6.1. Effects of Systems Building on Cost[a]

Country	Reported Reduction in Costs Because of Systems Building
U.S.S.R.	15%
Czechoslovakia	15%
Poland	15%
Sweden	None
Norway	None
Denmark	0–15%
West Germany	5–12%[a]
France	0–10%[a]
United Kingdom	5–10%[a]

Source: HUD, Industrialized Building, Comparative Analysis of European Experience, 1963.
[a] High rise only.

To interpret the figures in Table 6.1 it should be taken into consideration that these comparisons are made with conventional methods of construction, which in many parts of the world are still inefficient and antiquated, or where statistics are not totally reliable.

Interestingly, in countries where rationalized methods of construction are advanced, such as Germany or the United Kingdom, the economies are minimal. This is one of the reasons why systems building in the United States and Canada has not been as successful as in Europe.

Analysis of Cost Factors

A brief comparison of the traditional, but rationalized, construction methods with systems building will show the areas in which economies can be made and where costs increase. In a traditional system, the owner, who may also be the user, determines the needs of the user and commissions a designer who produces the design and specifications (see Figure 6.1).

A builder is then selected who subcontracts some of his or her duties. The builder coordinates workers and materials received from various sources to finish the construction.

Analysis of Cost Factors 157

Each stage of this process is more or less independent of others. Little attempt is made to interrelate the various activities or participants.

There is very little or no correlation between the activities of the owner to secure financing and, say, the work of subcontractors. Although some information may filter down through the system, the total process is not controlled.

In the systems building organization not only do other entities enter the process, but also there is a greater attempt to integrate as many as possible of these activities together as reflected in Figure 6.2.

Two new participants who enter the total process are system design and sponsorship.

System Design

This involves the complex process of studying the needs of the user; market analysis; development of standardized components; establishment of manufacturing and assembly layouts and process; allocation of resources and

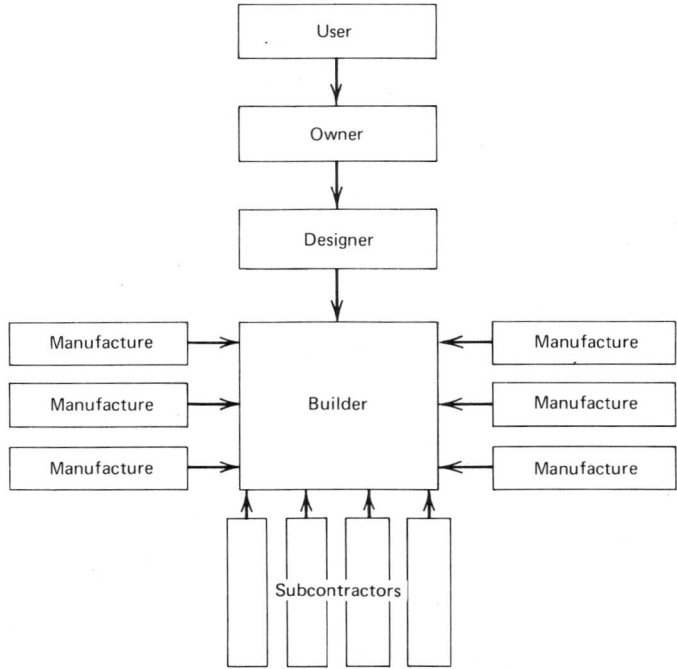

FIGURE 6.1. Traditional building diagram.

158   Building Systems Software

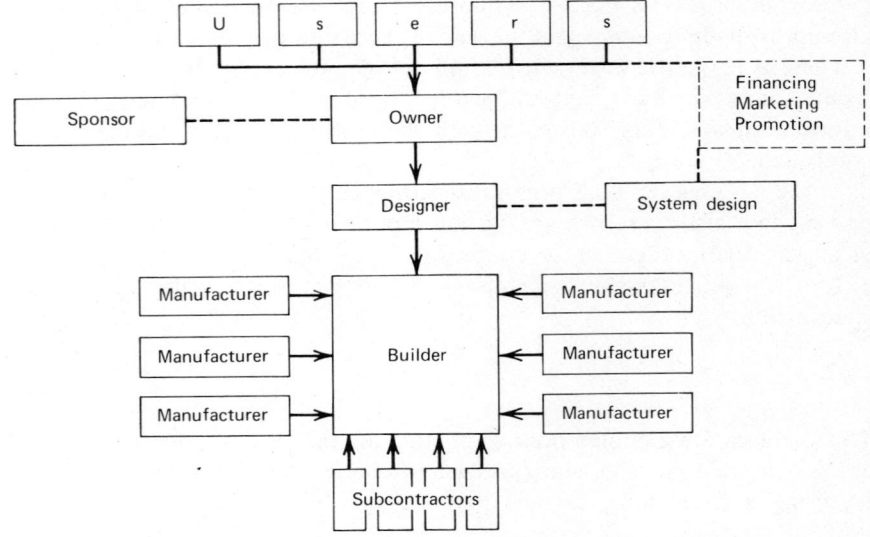

FIGURE 6.2. Systems building organization (schematic diagram).

materials and definition of a framework for the building designer to work within.

Sponsor

The role of sponsor is very important. It has been found that the major element for the success or failure of systems relies on the extent to which a sponsor (government, housing authorities, private enterprise) invests in the development, financing, marketing, and promotion of the system.

Naturally the owner, user, building designer, or manufacturer could play the role of the sponsor, but in most instances where government has been the sponsor, the success has been better assured.

A closer review of the systems building process shows that its total coordination revolves around (1) the employment of a variety of skilled persons to determine the extent to which industrialization and mechanization should be used and (2) the selection of the system—meeting the particular needs, resources, habits, and preferences of the user. Some of the skilled help needed include:

1. Sociologists to study the need of the users.
2. Planners to study the needs of the area.

3. Economists and financiers to determine the feasability and economic value of a given process.
4. System designers to provide the know-how in dealing with production processes, industrialization, and mechanization.
5. Architectural designers to produce functional and appealing systems for users.
6. Engineers to provide structural calculations and to ascertain the integrity of systems.
7. Managers to coordinate and correlate activities.
8. Administrators to oversee the disbursement of funds and the flow of materials and labor.
9. Marketing experts to study the markatability of the product.
10. Promotion experts to advertise and promote sales.
11. Real estate experts to facilitate and speed up sales.
12. Legal counselors for advice in all areas.

The influence of such a large organization on the final cost of the product is evident. On the technical side, the size of the investment (in equipment, installations, and materials) also has a strong influence on raising costs. Of course, there are many advantages in volume production. Since system building involves the volume buying of materials, the employment of unskilled labor, and higher factory productivity, great economies are made in these areas, compared to conventional methods of production.

The foregoing discussion reiterates the fact that the most important aspect of systems building is its higher productivity—and in some instances, the higher quality provided, which results in better value, but not lower cost. See Figure 6.3.

Other Software Aspects

The success of systems building largely depends on the software elements of the system rather than its physical nuts and bolts. Therefore, it is vital that the term "software elements" be understood as well its relative importance be recognized.

User

The most advocated and talked about software element, if it can be called that, is the user. The correct identification of the physical needs, preferences, aspirations, beliefs, and habits of the prospective user or users

160 *Building Systems Software*

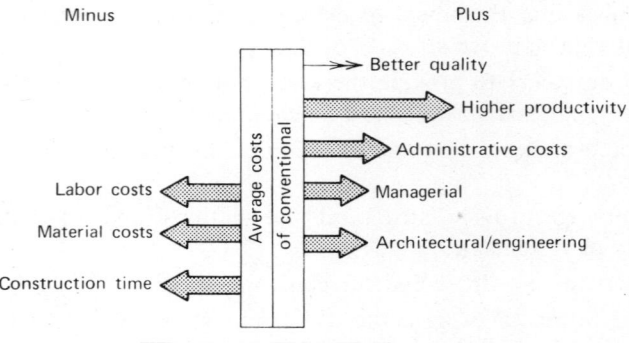

FIGURE 6.3. Schematic cost factors of systems building.

is particularly critical for the development of systems in a competitive, private market.

Many failures of systems building and traditional buildings are and have been attributed to the failure of the planner to comprehend the needs of the user. After all, the home is an extension of its inhabitant. It must respond to the needs of its inhabitants both in form and function. No single system is able to answer all needs or solve all shelter problems.

Owner

Unlike traditional construction, the owner cannot commission a unique design to be built. Systems must provide flexibilities that would allow more adaptability to users' needs.

If the owner is the sponsor, he or she should finance the operations of a socioeconomic study of the users' need, the marketing and promotional aspects and incorporate in its future plans a continuous feedback system, which will allow him or her, the owner, to continue the production of a marketable product.

Management

Management is required at various levels as follows:

1. At the decision-making level, the proper managerial skills are required to coordinate all activities.
2. At the production plant management is required to coordinate the proper flow of materials and products, scheduling, cost analysis.
3. At the site management is required to coordinate the workers, materials, equipment, and components and to interact and produce economical results.

Failure of managerial skills in any of these areas could have critical repercussions. It is particularly critical that at the level of the sponsor/owner, expert management be used. Unfortunately, this is an area in which little expertise is available around the world, and the exchange of ideas has been limited. The areas of marketing, production control, and planning also enter at this level of management.

Marketing

In the competitive societies marketing of a product is the essential ingredient for its sale. In the systems building process, this factor is even more essential. It is only through adequate marketing that a systems sponsor (or owner) may be able to have the assurance of a continuous demand.

Without the aggregated market, it would be foolish to invest in costly equipment and facilities to produce systems on an industrialized basis.

We are dealing with a very costly commodity that cannot be stored away in today's market prices; the components for the production of less than a 100 units will be worth more than $1 million. The investment costs of equipment and manufacturing facilities range between $1 and $7/8$ million for a factory producing 1000 to 1200 units. A factory that has its stock unsold for three weeks risks bankruptcy.

These are some of the reasons why marketing and market research are so essential in systems building.

Production Control and Coordination

The entire process of production takes on a complex format. Marketing assures the flow of orders; production fills out the orders. The coordination of the broad range of activities which lead to the delivery of product on time; the scheduling of production to run concurrently with the demand, the allocation of resources and materials to avoid delay or overproduction are clearly some essential functions.

Financial Control

The systems builder absorbs large expenditures in areas of equipment, installation, management, marketing, and production and must constantly balance the influence of these costs on the cost of the final product with a type of financial control to arrive not only at a competitive market price of the product, but also possible profits for the organization. This is probably the most important skill that the systems building management should possess—cost control.

Design

The task of the designer seems to be the most complicated and demanding. This professional must interpret the needs for a given use; concern himself or herself with the processes of fabrication, production, and assembly; consider the owner's requirements (profit); design a marketable and controllable product (production and financial); make all the cost-cutting decisions while preserving the quality and design values of his or her concept; and modify the design according to the feedback from the user, the market, and others.

Figure 6.4 describes graphically the interrelationship that should exist between the user, owner, designer, and builder to the sphere of software aspects within the systems building process.

The Question of Infrastructure

The foregoing factors relate to the software aspects from the production point of view. There are other factors that relate to the planning point of view without which full industrialization is not practical.

Systems building, as mentioned before, is the application of industrialized methods for design production, delivery, and the erection of buildings.

The industrialized process to be effective requires the existence of certain infrastructural amenities in the area, whether it be a city, region, state, or nation. Most of these amenities, despite having physical forms, are in fact considered as software elements, which may assure the success of systems building in a given area.

The most important area of such an infrastructure is the existence of master plans in the country or region that govern the usage of land through zoning laws, allow the establishment of codes and standards, permit the employment of available resources through judicious exploitation, assure the development of support industry to produce secondary components (doors, windows, toilet fixtures, cement, bricks, and so on), and establish laws and policies which encourage the industrialization of housing.

To this list we should add the necessary amenities to support industrialization: road systems, bridges, and overpasses and city utilities, including electricity networks, sewerages, water supplies.

The subject of infrastructure is too elaborate to be discussed here at length. However, it is a known fact that the government's role in the successful implementation of housing is an absolute must. In fact, a study of systems building around the world will show how the two are interrelated (see Figure 6.5).

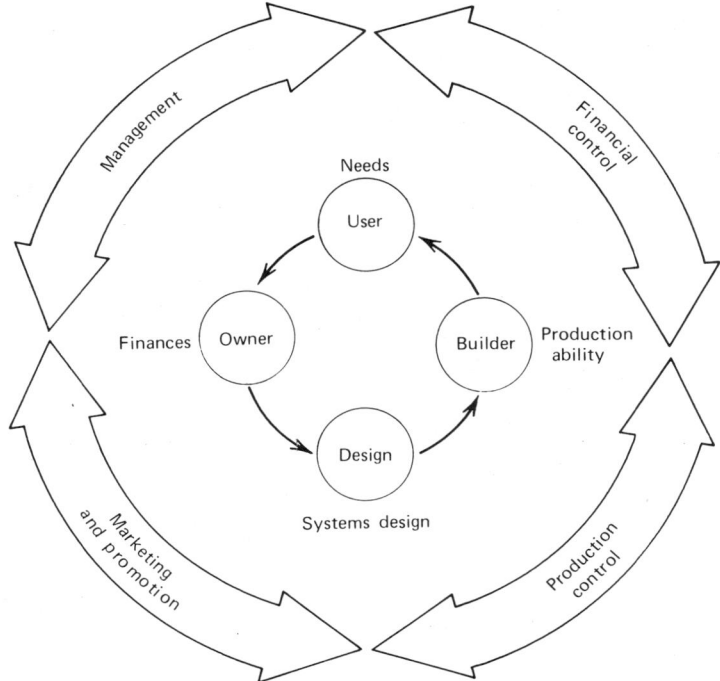

FIGURE 6.4. Software elements influencing the production level.

Most of the infrastructure required for development of systems building is also required for industrialization in general. Since this subject is very broad we will briefly describe only those areas directly affecting the production of systems building (see Figure 6.6).

Building Codes

Establishment of a uniform building code will assure the creation of building components of equal performance through a competitive market. This will have an important cost-saving incentive, while the structural strength and integrity of systems are controlled.

Standards

Standards are needed to provide minimum guidelines in areas such as control of quality and strength of materials; fire rating and sound control; impermeability, minimum room sizes, glazing, and sanitary needs. These

164   *Building Systems Software*

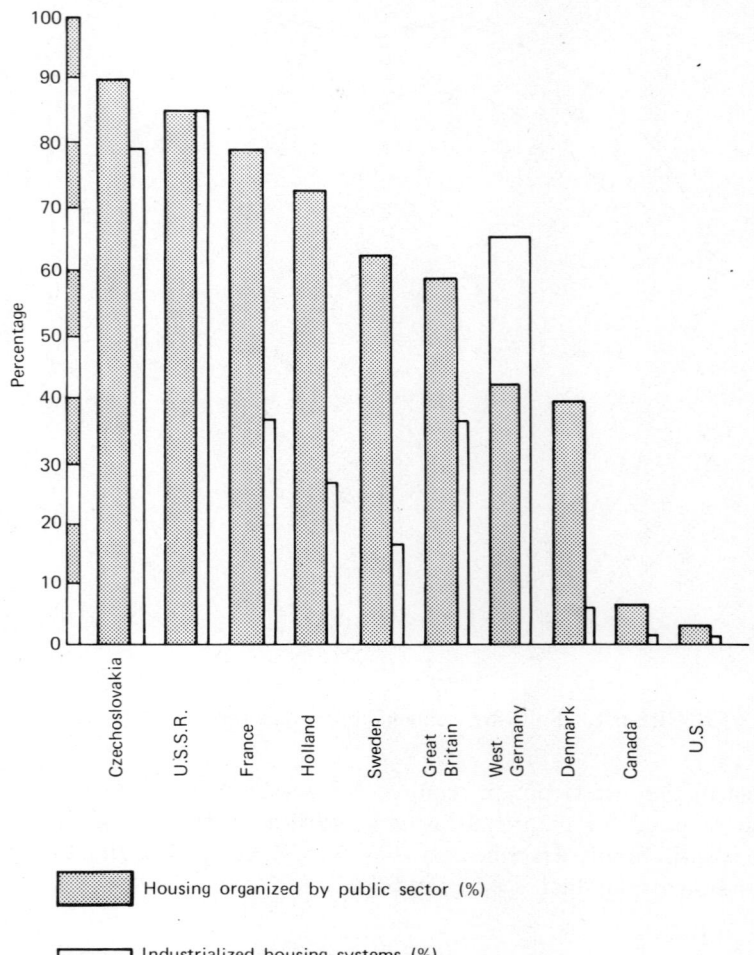

FIGURE 6.5.   Success of systems building depends on government involvement.

will assure that buildings respond to minimum requirements for safety and habitability according to parameters of current practice.

Zoning and Subdivision

Zoning and subdivision regulations seek to diffuse incompatible land use and control large-scale development by establishing criteria for neighborhood characteristics, growth patterns, and proportions of various building uses.

The Question of Infrastructure 165

Mortgage and Financing

The availability of a financing system for housing is an essential ingredient for its success. Housing represents the largest single expense in the life of a family. Any attempt toward making the costs absorbable will help to increase the marketability of the product. It is said that a 1% reduction in mortgage costs represents approximately a 10% reduction in the total costs of housing.

Any government action in the form of subsidies for low-income groups, the creation of a savings and loan system, the encouragement of housing cooperatives with special incentives, or other forms of taxation for the collection of money for low-income housing would improve the market for both standard construction as well as systems building.

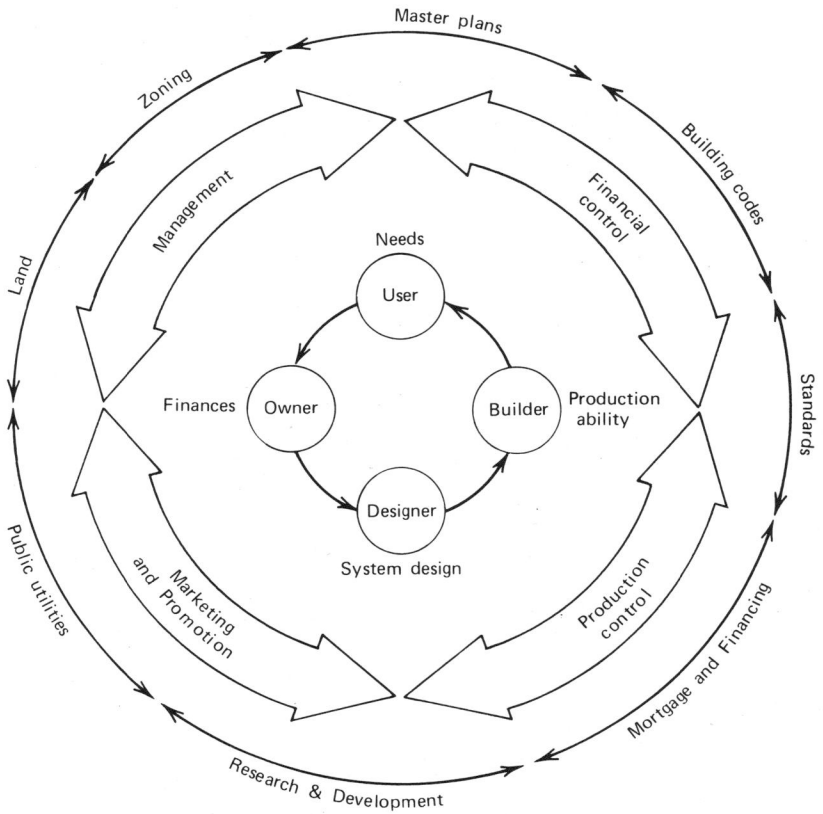

FIGURE 6.6. Software aspects of building.

Land

The availability of land for the purpose of development or the installation of systems is a very important factor. Statistics show that in urban areas, the cost of land has increased between 100% and 3000% during the last decade. The speculation on land is running without control. Many ideas have been proposed to alleviate this situation. Among the most serious ones are the nationalization of land and creation of land banks that lease the property to families for the duration of their stay in a given location. Methods of control could be devised through taxation, advance purchase systems, or price control.

Standardization and Modular Coordination

These two interrelated areas may have a serious impact on the reduction of costs in construction. The use of dimensional coordination makes the total construction process simpler and therefore less costly. Although the application of these principles imposes some limitations on the designer, most architects agree that standardization not only allows for creativity, it also gives more incentive for it.

Research and Development

One very important task of government is of course the investment in the research and development of new techniques, materials and methods, simplification of codes, implementation of better financing, and improvement of design methods—all of which lead to the establishment of national housing programs.

Decision Making and Systems Building

Altogether three levels of decision making lead to a successful systems building (see Figure 6.7). These represent the areas of macro decisions, which require government intervention; control at the level of production; and micro decisions at the level of the owner-user. The three levels must remain in constant interaction, relying on the support of each other.

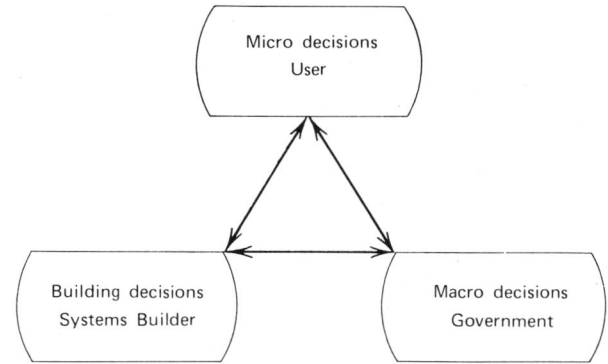

FIGURE 6.7. Three decision-making levels for systems building.

Conclusions

The technology to produce housing in factories is available today in the world. The developed nations have the physical capacity to produce housing at the rate of the demands. And certainly developing nations can import this competency and expertise if they desire to do so.

We can in fact produce homes like we produce automobiles or bicycles, but we are failing in our efforts because of a failure to cope with the software aspects of housing, which were discussed in this chapter.

The magnitude of the problem certainly requires the systems approach, but it also requires a lot of action at the level of public agencies and from the governments.

7

Systems Building Hardware

Iraj Majzub

Systems building is the process of producing buildings through the application of industrialized methods and management techniques for the purpose of increasing productivity and improving quality.

In the previous chapter we analyzed the software prerequisites to create the right surroundings for systems building to flourish. In this chapter we will look at the nuts and bolts of systems building and the most predominant proven systems existing around the world.

Classification

Typical classifications of industrialized construction divide these systems under three generic headings:

1. Frame or post and beam system.
2. Panel system.
3. Box or modular systems (shown in Fig. 7-1).

This classification, although descriptive of the major generic systems, in effect deals only with the structural aspect of the systems and needs to be expanded.

The Royal Institute of British Architects (RIBA) has proposed a better classification that makes a separation between the structural system and the infill subsystems. This classification although more explicit, is not totally

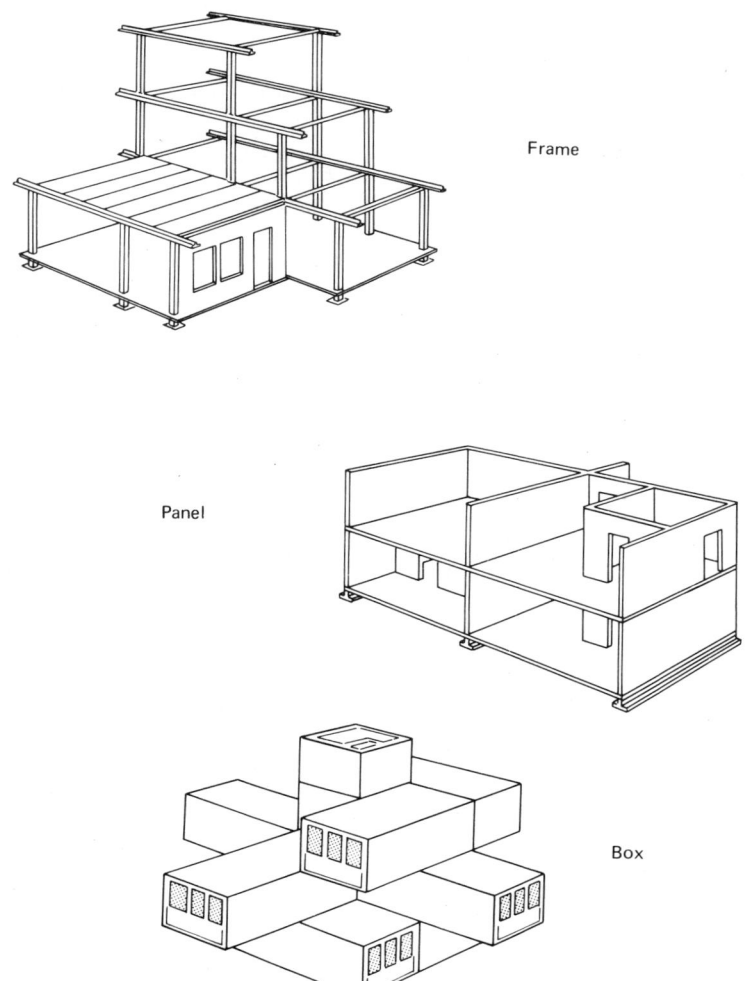

FIGURE 7.1. Classification of systems by their generic structural components.

clear-cut (see Figure 7.2). Systems, in fact, are not easily classifiable since each has characteristics that distinguish it from others.

To a great extent, the purpose of a classification is to facilitate the task of understanding the major constituents of the substance and to evaluate it.

In systems building the evaluation process depends on a large number of parameters that go beyond the structural aspects and deal with other

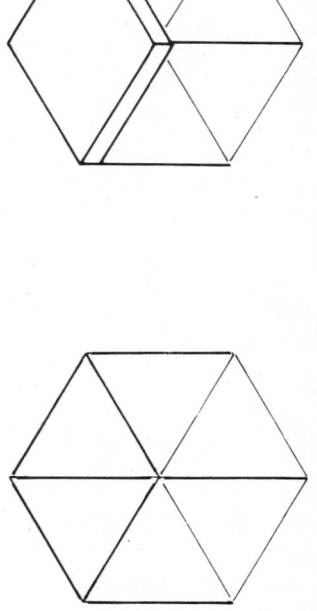
Structural frames when these are separated from other elements. Classification includes categories (28.1) for skeletal frames and (28.2) frame and slabs, both pin-jointed and rigid

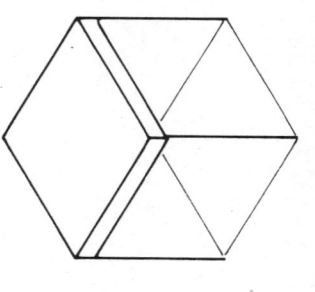
Slab structures for floors and roofs. Classification includes categories (23.2) monolithic and permanent formwork; (23.4) assembled plank and deck; and (28.9) space frames

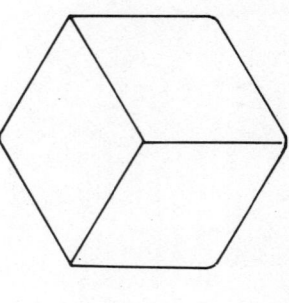
Structures within a total superstructure and not within one particular element. Classification includes (2–6) shell structures and spatial box and balloon structures

Structures supported on their external walls. Classification includes (21.1) for load-bearing walls, including cavity construction

Structures supported on their internal walls. Classification includes (22.1) for cross walls

FIGURE 7.2. Classification of systems building proposed by RIBA (England).

characteristics of the systems—such as its architectural features, the industrialized process used, the transportation and erection problems, and the socio-economic point of view. The following is a list of such desirable parameters:

STRUCTURAL
1. Durability.
2. Sound insulation.
3. Fire resistance.
4. Weather tightness.
5. Use of local available materials.
6. Maintenance free building.

ARCHITECTURAL
7. Flexibility in plan.
8. Flexibility in composition.
9. Adaptability.

DEGREES OF INDUSTRIALIZATION
10. Few components.
11. Integrated system.
12. High degree of mechanization.
13. Use of simple equipment.
14. Labor intensive versus capital intensive.
15. Need for unskilled labor.

TRANSPORTATION
16. Simple transportation package.
17. Economically transportable to long distances.
18. No need for special equipment.

ERECTION
19. Lightweight for easy handling.
20. No need for special equipment.
21. Rapid.
22. Low labor on site.

ECONOMIC AND SOCIAL
23. Lower cost for high quality.
24. Adaptability to needs of various socioeconomic groups.

To account for the foregoing desirable factors in a systems building, it seems that a classification based on the generic systems differentiation with subdivisions by the predominant weight of components would be more appropriate. See Figure 7.3.

The factor of weight influences the transportability of the components and has an effect on the production procedures and the erection onsite. These factors evidently influence the cost and make mass production less desirable or at least not feasible for many areas. The classification by weight also has the advantage of distinguishing between the various basic materials used in the production of components, which, by itself, could determine the characteristics of the system under study.

Generic Systems' Hardware

The length of this chapter will not allow us to enter into a detailed study of industrialized systems of construction. We will therefore discuss the main features of each generic system and show briefly the characteristics of principal subdivisions. For the purposes of a more rational evaluation of these systems, we will also describe the advantages and disadvantages of each in a brief table.

Frame Systems

Definitions

Frame structures may be defined as those structures that carry the loads through their beams and girders to columns and to the ground. The outstanding characteristic feature of such skeletal structures is therefore the reduction of the number and sizes of load-carrying membranes and a distinct separation between load-bearing and nonload-bearing elements in the building.

The modern skeleton structure, conventional or industrialized, is the result of the rational use of aluminum, steel, concrete, and wood in the building. The great strength of the new building materials and the acquired technology and engineering know-how of the last century allow us to go into larger spans and higher altitudes and thereby cater to the increasing demands of our modern age society.

Although the most standard frame structures are designed on an orthogonal grid system, frames of other kinds are also finding great application in the field of construction, such as geodesic dome structures and space frames.

Evidently, a frame structure by itself does not enclose the space, therefore infill elements are required to complete the system. Such infills may be built

Generic Systems	System	Production Materials
A: Frame systems	A1: Light frame	Wood Light gage metals
	A2: Medium weight frame	Metal Reinforced plastics Laminated wood
	A3: Heavy frame	Heavy steel concrete
B: Panel systems	B1: Lighweight and medium weight panel	Wood frame Metal frame (light gage) Composite materials
	B2: Heavy panel (factory produced)	Concrete
	B3: Heavy panel (tilt up-produced on site)	Concrete
C: Box systems (modules)	C1: Medium weight box (mobile)	Wood frame Light gage metal Composite
	C2: Medium weight box (sectional)	Wood frame Light gage metal Composite
	C3: Heavy box (factory produced)	Concrete
	C4: Heavy box (tunnel produced on site)	Concrete

Figure 7.3. Classification of building systems according to weight and form.

onsite or be in the form of prefabricated panels or both. It is also to be noted that since frames can remain independent of the infill structure, the system allows two distinct tendencies: (*a*) the skeleton is visible from outside, and (*b*) the skeleton is enclosed by a curtain wall system of similar panelized system, both offering great flexibility.

The idea of frame structures is not new. All through the Middle Ages buildings with skeletal frames were erected. In those instances the frame consisted of timber logs jointed together with masonry infill walls. In a sense, the old Egyptian, Greek, and Roman temples were attempts toward a frame structure.

In the field of housing, this type of system has had less application. However, in larger scale structures, such as in institutional buildings, very advanced designs have been produced in frame systems. The advantage of frame systems is mainly in the interior flexibility that the use of the frame structure offers: large spaces can be allocated to the different functions required by the institutions and subsequently modified or changed around according to need. A well-coordinated frame system such as SCSD (The School Construction System Development) in the United States and SEF (Study of Educational Facilities) in Canada are designed to allow this flexibility not only in plan, but also in height.

Although the concept of frame systems is not very far from that of the traditional frame structures, it is the most apt to the idea of personalization and other innovative ideas in the field of housing.

Light Frame Systems (A1)

These systems use wood or light gage metal studs or posts for the construction of the frame. The balloon frame or Western framing construction in North America (see Figure 7-7) or similar systems using aluminum and light gage metal as well as wood post and beam construction and precut cedar homes are examples of such systems. This system is not much different from the conventional methods of construction known in North America (see Figure 7.4) and has influenced the conventional market to the extent that the buildings of this type are now considered rationalized construction. See Table 7.1, Figures 7.5 to 7.9.

Medium Weight Frame Systems (A2)

In the field of housing this type of system has had little application. However, in larger scale structures, such as schools, very advanced designs have been produced. The best known of such examples are the SCSD of California, the SEF of Toronto (Canada), and CLASP (Consortium of Local Authorities Special Program) of England (see Figures 7.10 to 7.13).

The advantage of these systems is mainly in the interior flexibility that the use of the large frame structure offers; large spaces can be allocated to the different functions required by the institutions and subsequently modified or changed around according to need. A well-coordinated frame system is designed to allow this flexibility not only in plan, but also in the height of rooms and the interchangeability of subsystems, such as partitions, exterior wall, light fixture and HVAC. See Table 7.1

Heavy Frame Systems (A3)

These are built in reinforced concrete. The application in the field of housing has been rather limited ("Componoform" and "Mitchel System"). In institutional buildings, RAS in Montreal, Canada, is a good example. The major application of such systems is in industrial buildings and office buildings. (Projekton, Holland) but, the most successful heavy frame systems are the prestressed concrete components such as Spancrete (Canada)—Durastress (U.S.) which have had application in all types of buildings. (Figures 7.14 and 7.15)

The basic drawback of concrete frames is in the jointing process of columns, beams, and girders. It is very hard to produce continuity in the structure without having heavily built joints. This factor obliges the designers to go into large size columns and beams and, therefore, larger spans. Hence, the creation of mega structures or concrete arches to reduce the number of connections.

An example of a mega structure in concrete can be found in the Townland project for HUD's Operation Breakthrough. The column-to-column connection and the column-beam attachments are staggered to allow for easier access to individual joints.

The technology is essentially that of a parking garage with longer spans. An important difference, however, is that the super frame has long spans in both horizontal and vertical directions, the latter properly known as long column design. The system spans between 30 and 60 ft. See Table 7.2.

TABLE 7.1. Performance Evaluation: Light and Medium Frame Systems (A1 and A2)

	+ Advantage	=	− Disadvantage
Structure	Lightweight components, structural support in frame	Life span 25–50 years	Sophisticated jointing; connection problems
Architectural application	Flexible use; large open spaces; modifiable; evolutionary	Adaptable to various sites	Limited application to housing
Industrialized process	Simple tooling; simple problem of fabrication; rapid; use of semiskilled labor	Average capital outlay 20–30% industrialized	Average degree of standardization; closed system (20–30%)
Transportation	Easily transportable, stackable parts; demountable, no limit of transport		Many components; may get misplaced
Erection and assembly	Simple equipment needs except on high rise		Labor intensive, costly; time-consuming; skilled personnel

176 Systems Building Hardware

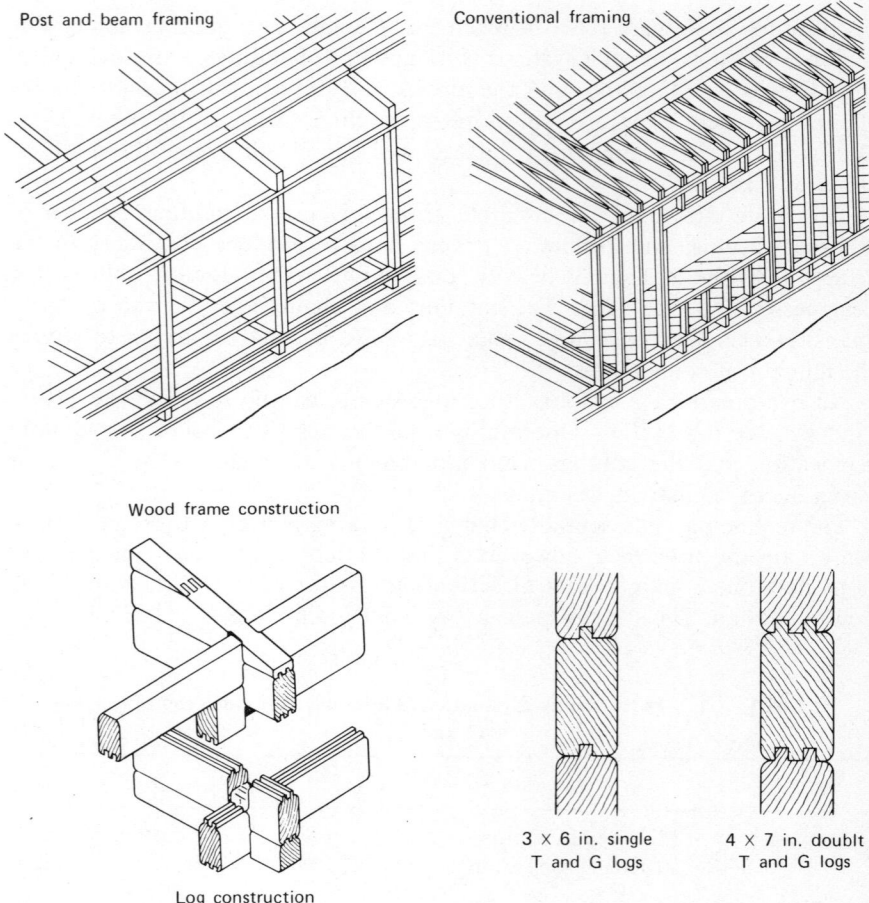

FIGURE 7.4. Conventional wood frame construction is the predominant type of rationalized construction used in North America for residential use (Photo courtesy of Automated Building Components, Miami, Florida).

Panel Systems

Definition

Panel systems may be defined as those structures that carry the loads through large floor and wall panels. There is a tendency to confuse panel systems with what is predominantly used in this field, namely, the heavy concrete panel system. This, of course, is not correct. Panels are made in a

Panel Systems 177

FIGURE 7.4. (Continued)

variety of forms and materials and are erected onsite together to form the final unit. Concrete panel systems have been extensively used in Europe in the production of high-rise buildings, while other panel systems (in wood, plastics, light metal, and wood frame) have been primarily used for low-rise buildings.

Classification

In this chapter we will examine the existing panel systems from the standpoint of the weight of components as follows:

1. Lightweight panels (wood frame, papercore, and plastics).

FIGURE 7.5. Production of trusses through the use of gang nails and special roller (Photo courtesy of Automated Building Components).

TABLE 7.2. Performance Evaluation: Frame Systems (Heavy) (A3)

	+ Advantage	=	− Disadvantage
Structure	Structural support in frame; fireproof and soundproof	Life span 75 years	Heavy weight; complicated joints; many components
Architectural application	Flexible plans; large open spaces; modifiable interiors	Adaptable to various sites	Application to housing only in mega structures; nondemountable
Industrialized process	High degree of industrialization	Average capital outlay 20–30%; industrialized	Heavy trucks and handling; costly transportation; limit of transport
Erection and assembly	Quite rapid	Average time outlay	Heavy equipment; labor intensive; skilled personnel

2. Medium weight panel systems (lightweight concrete or steel frames).
3. Heavy weight panel systems (reinforced concrete panels).

The separation by weight of components is not entirely arbitrary as the weight (therefore, size) has a direct effect on transportation and erection problems, jointing, handling equipment, and the height of the building.

Panel systems are not a totally industrialized system since they provide the enclosure of the building, and the finishing or addition of mechanical-electrical systems are generally extra. They are only 40 to 60% factory finished dependent on the degree of sophistication of the system.

In a panel system there are many compenents (15 to 30 per average unit). Therefore, more joint treatment and onsite labor are required.

Handling of panels is relatively simple when they are built of lightweight materials. Therefore, simple handling equipment is required. In heavy panel systems the weight becomes a major handicap since it influences the size of

FIGURE 7.6. A gang-nail detail (Photo courtesy of Automated Building Components).

180 Systems Building Hardware

FIGURE 7.7. The use of premanufactured trusses increases the speed of production in wood frame construction (Photo courtesy of Automated Building Components).

the panels and their possible span. This results in monotony in the design. Naturally, the weight also has an influence on the height of buildings.

Despite these considerations, panel systems (particularly in medium lightweight) represent a breakthrough in the field of industrialized housing. They increase the onsite productivity and allow for a much faster completion of the units. The result is a more economical package.

Light & Medium Weight Panels (B1)

Lightweight Sandwich Panels

These panels are basically of a standard size and produce both the walls and roof of the housing units. They are produced in sandwich form and use two face sheets and a lightweight core bonded together in a factory.

Panelfab and Prepsa (Miami, Florida) in general use a honeycomb craftpaper core metal face sheets. (See Figure 7.16).

As a result of Operation Breakthrough there were some new developments in this process. TRW produced a panel incorporating more sound and thermal-absorbing qualities (see Figure 7.17). Ball Brothers proposed a sandwich panel that used urethane foam as a core and fibrous materials as face sheets; Materials Systems of California used recycled materials to produce a corrugated core and face sheets (see Figure 7.18).

All these systems have the advantage of lightweight and ease of erection. However, they all have serious connection problems. See Table 7.3.

Wood frame Panel Systems (Medium Weight)

In North America, panelized construction is synonymous with wood panel packaged housing. A survey conducted in January 1975 showed that out of 951 builders interviewed, 61.9% were currently using panelized wood packages.

The major reasons why builders used panelized construction were found to be (in order of preference): (1) fixed price of materials, (2) faster completion, (3) simplified purchasing, (4) ease of year-round construction, (5) faster money turnover, (6) less complicated, (7) reduced pilferage, (8) better quality, and (9) reduced equipment investment.

FIGURE 7.8. The use of premanufactured trusses increases the speed of production in wood frame construction (Photo courtesy of Automated Building Components).

FIGURE 7.9. Lightweight metal frames that prefabricated and shipped to site (Photo courtesy of U.S. Steel).

The equipment required for production of the panels is very simple and may not exceed that of carpenter tools, although efforts toward the mechanization of production has resulted in some machinery capable of increasing productivity and reducing the labor requirements. These include machines such as automatic nailers, hydraulic presses, and rollers for production of trusses; sheathers; and automatic saws capable of precision cutting of any length of studs or angle cutting.

There are a large number of producers of panelized packages in North America and elsewhere in the world. While the producers use very similar techniques for the production of panels, they may have major differences in their marketing, management, financing, and transportation methods.

TABLE 7.3. Performance Evaluation: Panel System (Light) (B1)

	+ Advantage	=	− Disadvantage
Structure	Structural support in frame; lightweight	30-year life span	Jointing problems
Architectural application	Good in housing; expandable	Minimum finish onsite; mediocre space arrangements	Low rise only; limitations in span; rigid use of space
Industrialized process	High degree of industrialization in production of panels	Medium capital; outlay average; cost of tooling 50% industrialized	Need for local finish, wiring, plumbing
Transportation	No limit to distance; simple; inexpensive	Several components	Joint problems
Errection and assembly	Easy handling; simple assembly process		Loss of components; problems of mismatch

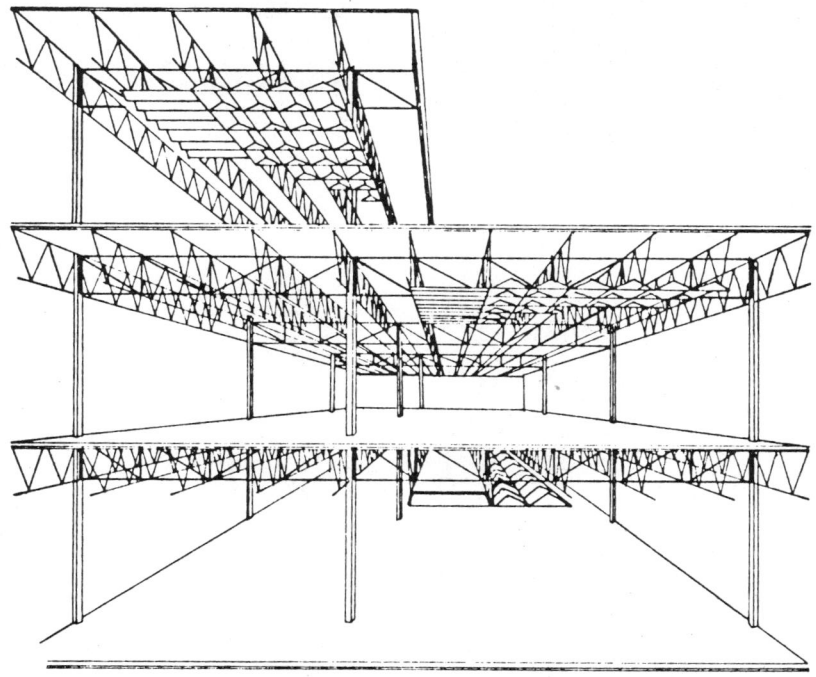

FIGURE 7.10. SEF system, Canada: This system is to a large extent an open system that allows the integration of all subsystems according to precisely laid down performance criteria.

FIGURE 7.11. A joint detail, SEF system (Photo courtesy of CMHC, Ottawa, Canada.)

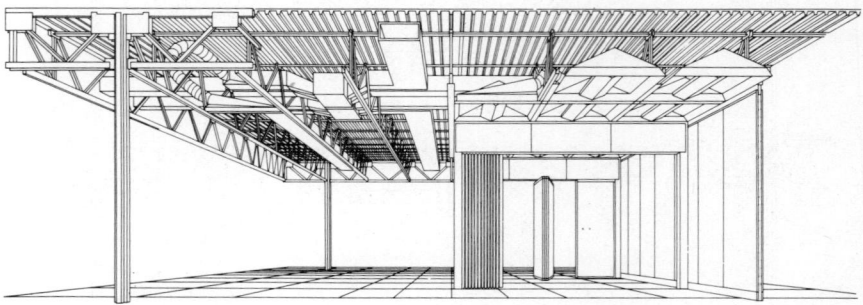

FIGURE 7.12. SCSD, California (USA): A totally integrated system of coordinated components. Note that the area between ceiling and roof deck is designed to allow maximum flexibility in usage of subsystems: plumbing, lighting HVAC, and so on.

Light & Medium Weight Panels 185

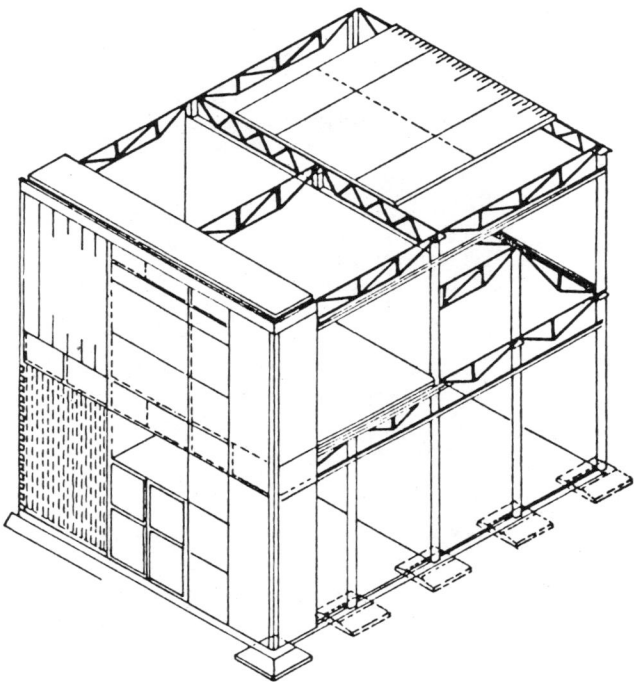

FIGURE 7.13. CLASP system (United Kingdom): A set of coordinated components providing the structure and the main enclosures of the building.

One of the major leaders in the field of package housing and lightweight panel systems, wood frame, in North America is the National Homes of Lafayette, Indiana. This company has been one of the few that has survived the fluctuations of the market during the last 35 years and is still producing a large number of units through its various industrialized systems.

As mentioned earlier, the success of this kind of operation is basically because of the marketing and management capabilities of these corporations, rather than any major technical innovations and technology of production. The small prefabricator and the giants are primarily using the same simple technology and offer preplanned, precut, or preengineered packages to builders.

To offer a variety of plans, home plans are broken down into components, each of which is numbered and preplanned to the smallest subassembly. (See figures 7.19 and 7.20) These components are repeated in as many models as possible to bring the number of components down. Every conceivable detail is worked out in advance and the builders get free

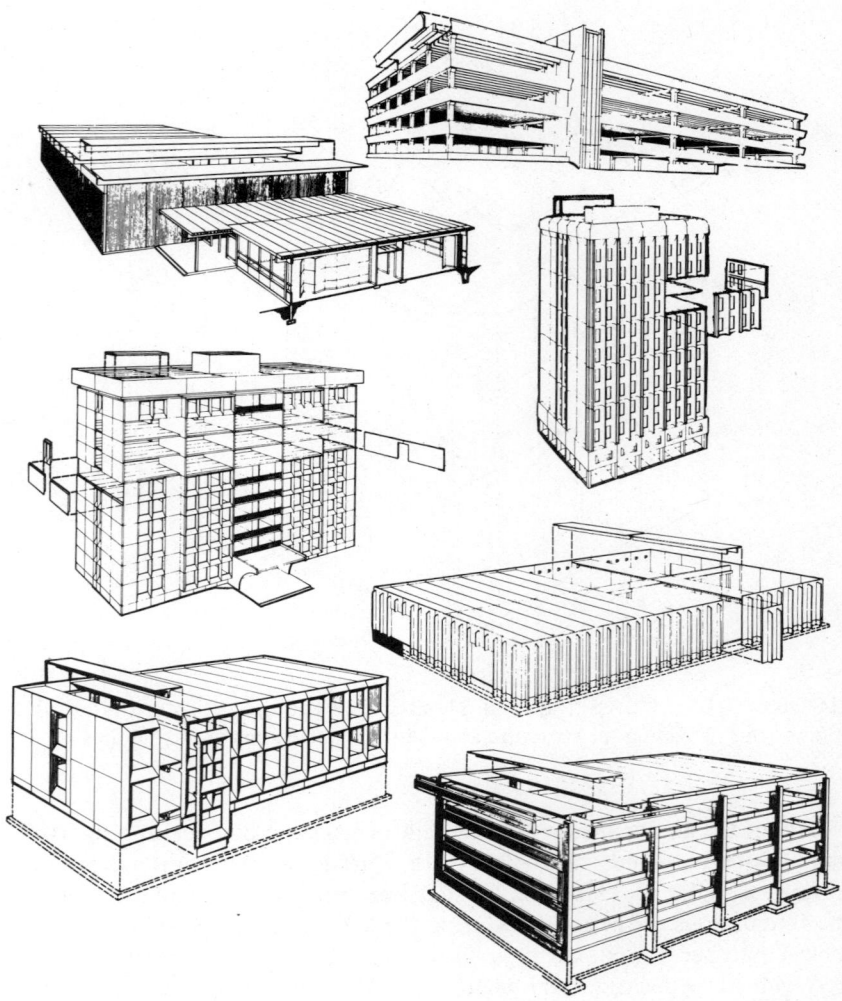

FIGURE 7.14. Various applications of prestressed concrete components (Courtesy of Dura Stress Inc., Leesburg, Florida).

advice for the assembly of the components as part of the package. See Table 7.4.

Heavy Panel Systems (B2)

Historically, Europeans popularized the use of concrete panel construction with the development of several sophisticated systems for the production of

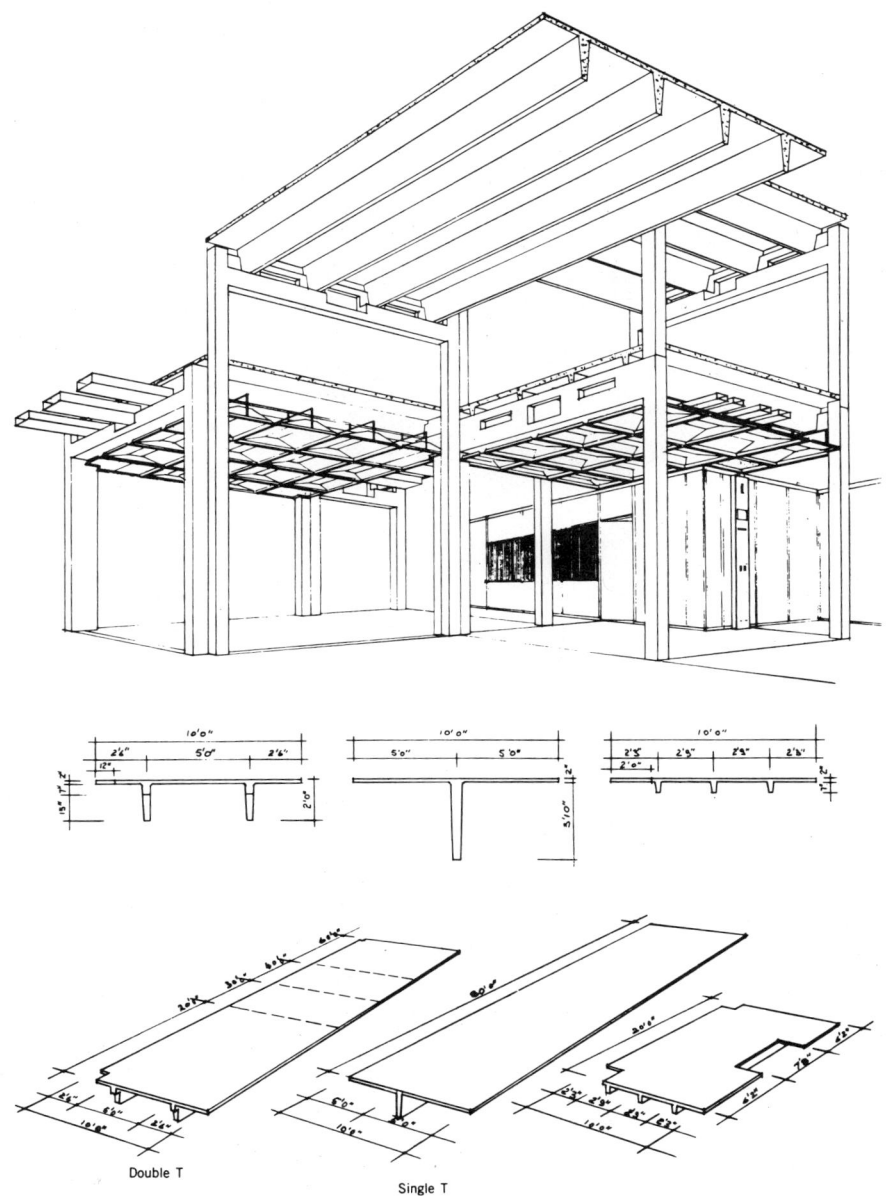

FIGURE 7.15. General perspective view and some components of RAS System produced by Francon Group in Montreal. Note the similarity of this system with the prestressed concrete components available on the open market in North America.

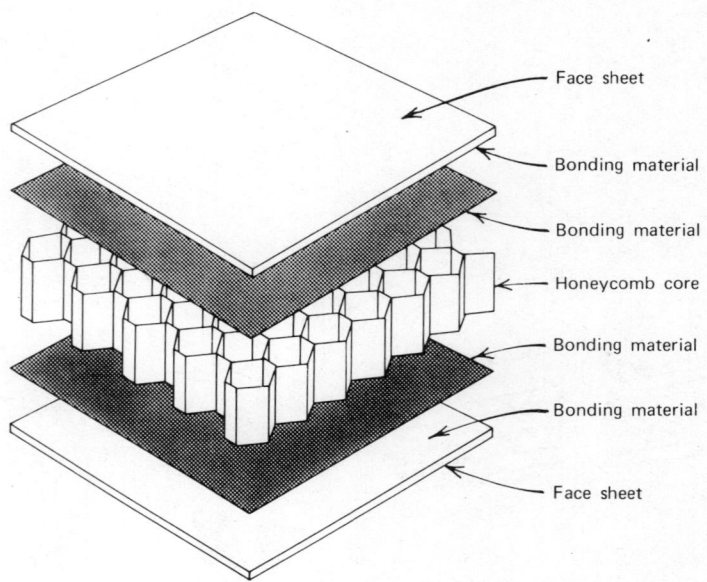

FIGURE 7.16. Basic components of lightweight honeycomb sandwich panels.

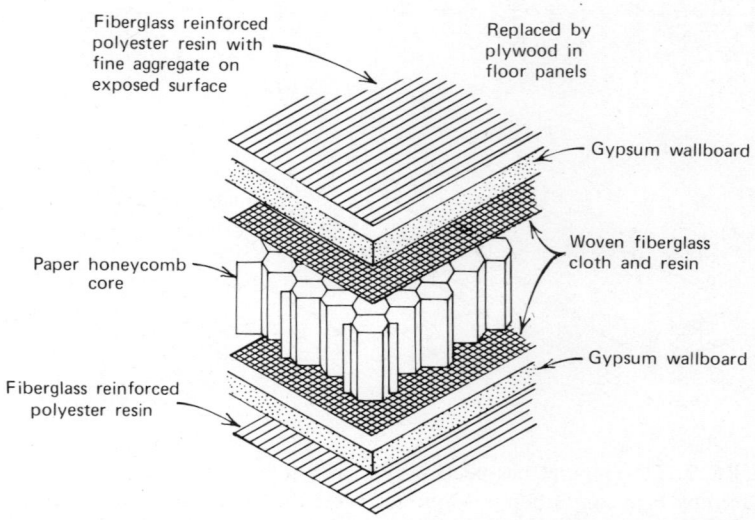

FIGURE 7.17. The sandwich panel, TRW system, Operation Breakthrough.

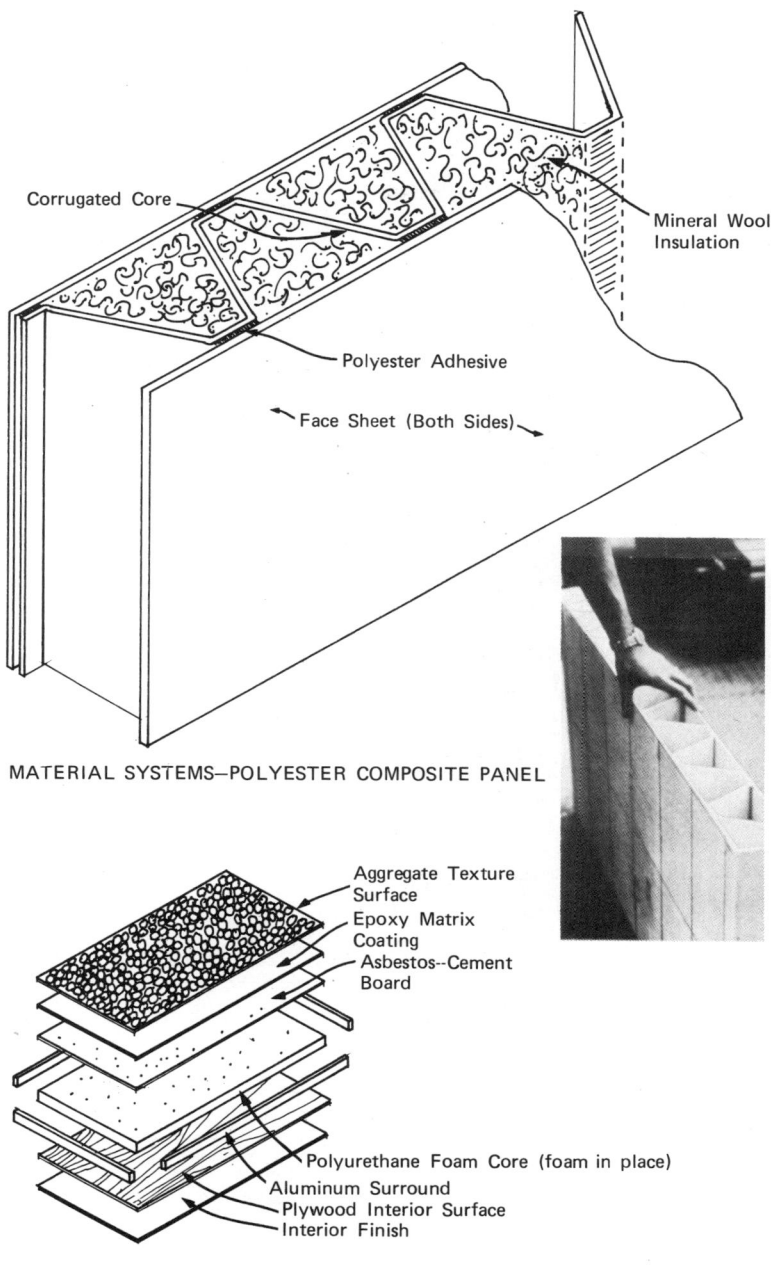

FIGURE 7.18. Two other lightweight and medium weight panels produced for Operation Breakthrough. Above: Material Systems Composite Panel. Below: Pantek Sandwich Panel Construction, produced by Ball Brothers (Courtesy HUD, Washington, D.C.).

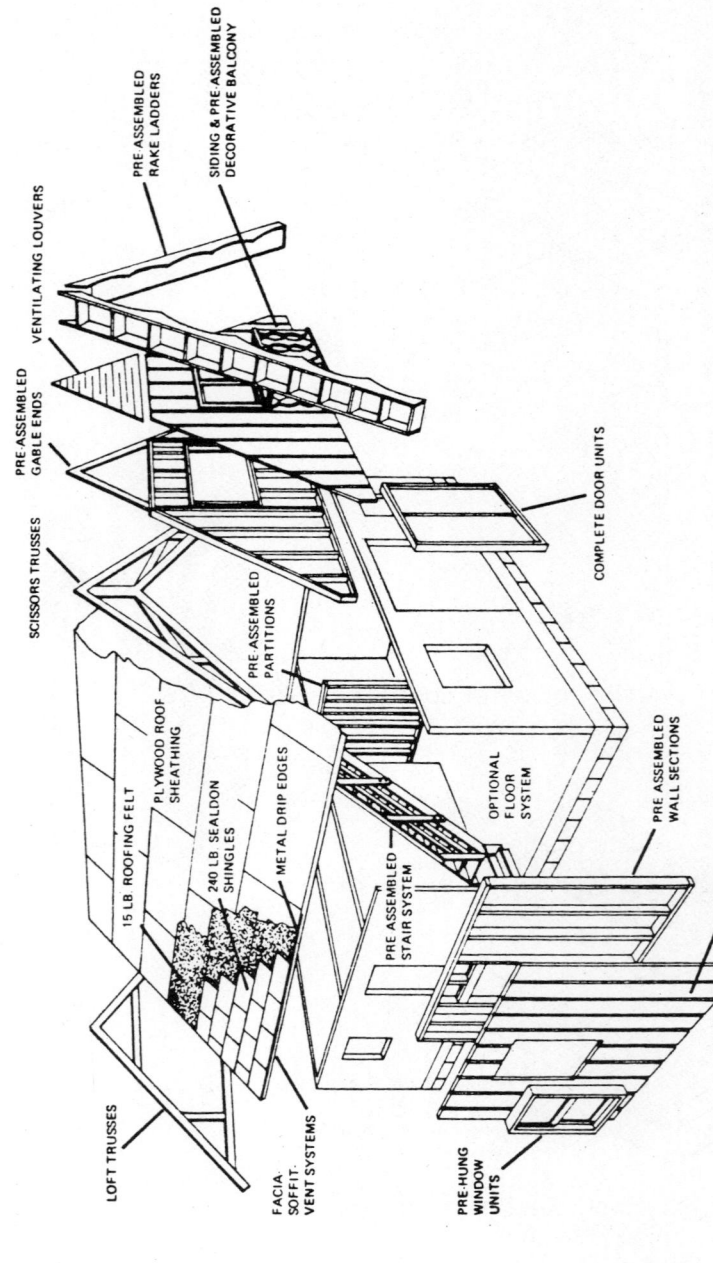

FIGURE 7.19. A typical package of components provided by the Wickes Corporation's "quick & easy" preassembled vacation home (Courtesy Wickes Corporation, Saginaw, Michigan).

FIGURE 7.19. (Continued)

hundreds of thousands of residential units after the World War II. However, the idea of prefabrication in concrete panels has its origins in the early 1900s and in the development of casting molds for concrete tilt-up panels in the United States.

From the mechanization point of view, the invention of reinforced concrete in the early nineteenth century and its theoretical as well as experimental development during several decades was one of the most important innovations brought about in the field of construction. This was the first artificially built monolitic material derived from the union of concrete and steel that can take any form given to it by the mold.

Postwar Housing Conditions

World War II had, of course, brought about enormous amounts of destruction in Europe, but the shortage of housing was not the only reason for resorting to industrialization in housing and the use of concrete panels. There were other factors, including the scarcity of labor and material, the

TABLE 7.4. Performance Evaluation: Panel System (Medium Weight) (B1)

	+ Advantage	=	− Disadvantage
Structure	Lightweight load-bearing components; easily erectable; structural support in panel	Simple jointing; to be fire proofed; average life span 25–30 years	Many components; jointing problems
Architectural application	Good application in housing; personalization	Flexible usage of space	Low-rise building
Industrialized process	Simple process of fabrication-low capital outlay; labor intensive; semiskilled and unskilled labor; simple tooling	Medium degree of standardization (50%); applicable on national scale	Low degree of industrialization; finish onsite plumbing and electricity
Transportation	Simple; long range	Limit of transport 1000 miles and more	Many components; damage
Erection and assembly	Easy handling; simple assembly process	Relatively simple erection; no need to handle equipment on low rise; semiskilled personnel	Loss of components during assembly; joint problems

virtual standstill of the building industry, the antiquated existing construction techniques, the demand for better standards of living as a result of rising affluence, an increase in household formations, migration to large urban centers, and a growth in the number of births.

The following examples will show the magnitude of the problem. (In Europe, including the Union of Soviet Socialist Republics, some 44 million housing units were either completely destroyed or badly damaged):

1. Yugoslavia lost 25% of its housing; Poland, 22%; Greece, 21%; Russia, 20 to 30%.
2. Wartime destruction was equal to some 15 years of the prewar rate of building.

3. The time of conventional construction of a typical house in France was about 2500 hours in 1956; by 1966 it was reduced to half that amount through industrialization.
4. All European countries had a low rate of yearly production of housing.
5. Labor was in short supply, and instead of unemployment, workers had to be imported. It was difficult to employ cheap labor.
6. Land became an expensive commodity and in short supply, particularly around metropolitan areas.

Government Intervention

The enormity of these problems obliged the European governments to take an active part in the field of construction and prepare large-scale housing programs, which are probably the most important factors in the development of systems building.

FIGURE 7.20. Application of a medium weight panel system in geodesic dome structures (Courtesy Geodesic Structures, Inc., New Jersey).

Contracts of 3 to 5 years for the construction of thousands of units were given out to manufacturers of systems. It should be noted here that the concrete panel systems were selected in Europe primarily because they were able to increase the productivity with the use of a smaller labor force. The purpose was not cost reduction, since experience has shown that this has rarely been achieved.

The concrete panel system is built to last for many years, having the advantage of being soundproof and fireproof. The system used in the manufacture of the panels is relatively simple in technique, but highly capital intensive. It employs only skilled and semiskilled labor, both for the manufacture of the panels and for their erection on site.

Concrete wall panels are produced by highly sophisticated machines that require a bare minimum in manpower. The machines are easy to adjust and shop-trained workers can easily master their manipulation. New techniques and aggregates are used to make larger components, with the weight of them being under control. In some systems, the mechanical and electrical subsystems are preassembled and are integrated in the panel during casting.

In the Western European countries, six or seven major companies build approximately one half of the total panel systems. Each company owns many international licenses and has large production capabilities. These companies are Larsen & Nielsen (Denmark); Skarne (Sweden), Balency (France), Jesperson (Sweden), Tracoba (France). (See figures 7.21 to 7.31). Similarly in Eastern Europe, heavy panel systems are used for production of the large majority of public housing. (See Fig. 7.32).

North American licensees of panel systems and similar local systems have not had great success on this continent, particularly where low- and medium-cost housing were envisaged. (Figure 7.33). Systems such as Balency, Rose Wates, and Tracoba were tested and built by Operation Breakthrough and did not show any economies in cost comparable to the cost of high rises built with rationalized conventional methods. Only a continuous high volume in demand would make such systems feasible in low-cost housing—and this only through public or cooperative housing.

Since in the concrete panel systems the weight is a limiting factor to the size and span of the concrete, the panel system in general looks very repetitious and monotonous unless features such as additional balconies are added to improve the exterior outlook. The floor layouts also remain rather limited.

Heavy Panel Systems (Tilt-Up Concrete Panel System) (B3)

An economical method of onsite panel production is the tilt-up concrete wall system. Tilt-up construction is a special form of precast construction generally considered to be a new technique since most of the buildings

Government Intervention   195

FIGURE 7.21. Balency system used in Scotland Thamesmead Development (Courtesy Impresa MBM s.p.a., Milano, Italy).

erected by this method have been built after the 1950s; however, examples of tilt-up systems were used as far back as 1908 in the United States. Tilt-up buildings up to 8 stories high have been erected, although this method is generally used for erection of 1-story buildings. See Table 7.5.

Fabrication

Tilt-up panels are generally poured on the floor slab of the building to be erected or on external platforms prepared for this purpose. The floor must be level and smooth and openings temporarily closed. To obtain a uniform floor slab, a bond breaking agent is applied to the slab.

Panels can be poured in all sizes, and they may be given a variety of surface finishes with the use of simple techniques. The thickness of panels is generally between 4 and 6 in. and according to the need is reinforced and insulated.

Since the floor slab is used as the "form," very little time and material is spent on preparation of the "edge forms," which are usually made of 2 in. lumber at the required height. The lumber is perforated at intervals of 18 to

FIGURE 7.22. Although heavy panel systems are primarily used in high-rise construction, they recently started being used in Europe in low-rise residential construction (Photos courtesy of Raymond Camus & Co., Paris, France).

24 in. to allow the passing of reinforcement rods. Small wooden or steel saddles are used to keep the reinforcing in the desired position. These are removed as the concrete is poured in the forms. Forms for openings and windows are positioned on the floor and secured by use of weight. Window sashes can be incorporated in the panels.

In this type of construction it is important to carefully work out a schedule of operation. Tilt-up panels may be built to be load bearing or act as infill panels. The jointing of panels, therefore, may be achieved in a variety of ways.

Once the panels are cured, they are lifted up with the help of cranes or other lifting devices and set on the foundation wall. The jointing of walls to the foundation wall is generally done with the help of Portland cement

TABLE 7.5. Performance Evaluation: Panel System (Heavy) Concrete (B2)

	+ Advantage	=	− Disadvantage
Structure	Load-bearing components; durable; soundproof and fireproof	Average erection time; life span 50–75 years	Heavy weight components; unrenewable resources; jointing problems
Architectural application	Good for housing (medium and high rise)	Some finishing onsite	Rigid usage of space; limit in size; span permanent
Industrialized process	Economical on large-scale production	Good degree of standardization (50–60%)	Capital intensive; expensive tooling; need for onsite finish grouting, plumbing, and HVAC connection problems
Transportation		Several components	Costly to transport; limits on size and weight; special machinery; limit of transport 100 miles
Erection and assembly	Simple erection procedure	Average time	Costly handling equipment; skilled labor; jointing alignments problems

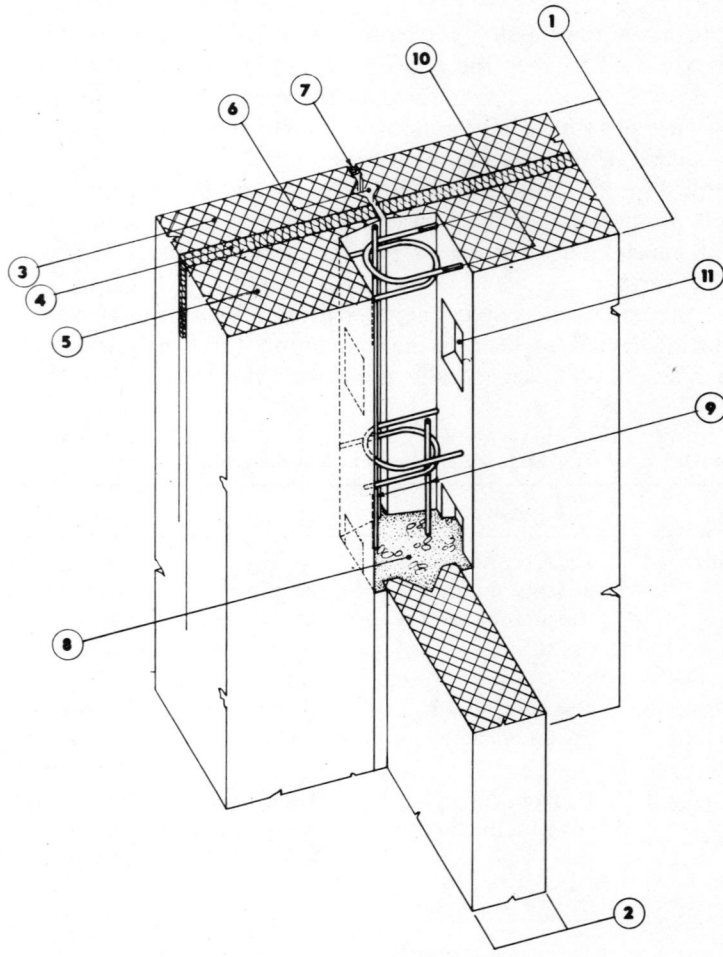

FIGURE 7.23. Shear transmitting vertical joint, isometric view. 1. Exterior panel 2. Interior panel 3. Outer layer 4. Rigid insulation 5. Inner loadbearing layer 6. Expansion and drain chamber 7. Sealing 8. Concrete filling 9. Vertical key steel bars 10. Horizontal steel bars for resistance to tensile stresses 11. Pockets for resistance to comprehensive stresses (Courtesy of MBM s.p.a., Milano, Italy).

mortar, but premolded joint filler has also been used. Another procedure is to set the panel on pads and then fill the joint with mortar dry-packed into place.

When panels are acting as infills and columns are used, the column is generally poured after the panels are set in place; therefore, the jointing problem is simplified.

The major area of application of tilt-up construction has been in large industrial buildings, but several attempts have been made to popularize this method in the construction of housing and apartment houses. (Figure 7.34). Altogether, the tilt-up construction seems to be more economical than other types of heavy panel systems since it eliminates the shipping and transportation costs and some required handling costs. (See Table 7.6.)

Box Systems

Definition

Box systems may be defined as those systems that use tridimensional modules (or boxes) for the fabrication of habitable units. The major characteristic or such modular units is in their internal stability or the fact that the box can withstand loads from various directions. This is a great asset in industrialization since it allows the transportation or handling of large prefabricated sections at one time, which in turn allows a higher degree of factory control and finish of the product and reduces the number of man-hours required for the assembly onsite. From this point of view, therefore, boxes are the more sophisticated and advanced system within the framework of systems building.

The degree of industrialization achieved in box systems may vary, depending on the type of box used. The most finished factory-built box is, of

TABLE 7.6. Performance Evaluation: Panel Systems (Tilt up) (B3)

	+Advantage	=	−Disadvantage
Structure	Load-bearing wall; durable; fireproof and soundproof	Life span 50–70 years	Not suitable for high rise
Architectural application	Good for lower-cost homes	Good usage of space; some finish on-site	Limited usage in housing; limited flexibility; permanent nonmodifiable
Industrialized process	Economical; labor intensive, (unskilled); economic tooling	Degree of mechanization 20–30%	Slower than factory built; need for local casting of columns, roofing, interior finish, and so on
Transportation	No problem		
Erection and assembly	Simple process	Average time for erection	Need for a crane; scheduling problems

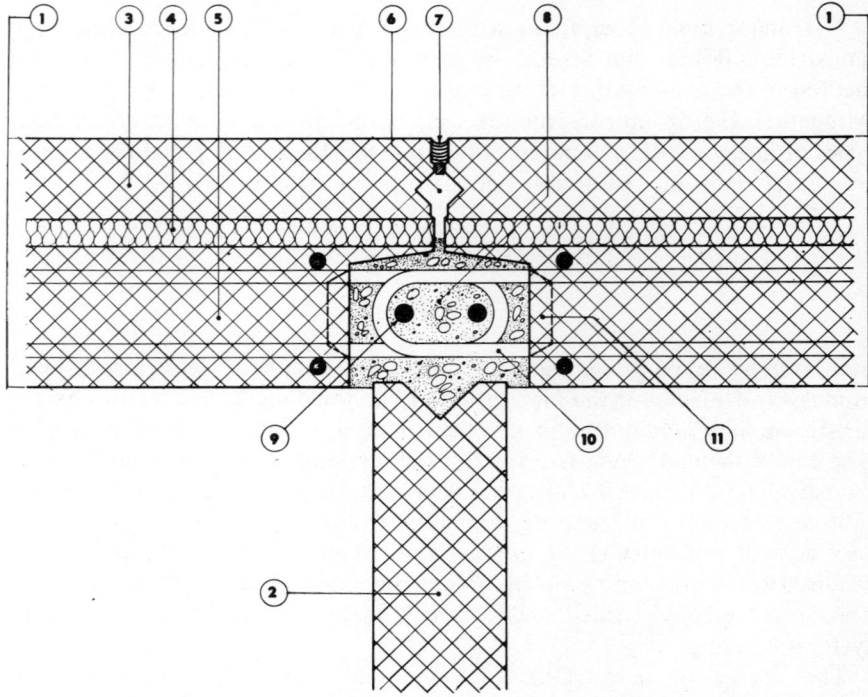

FIGURE 7.24. Shear transmitting vertical joint, horizontal section, 1. Exterior panels 2. Interior panel 3. Outer layer 4. Rigid insulation 5. Inner load-bearing layer 6. Expansion and drain chamber 7. Sealing 8. Concrete filling 9. Vertical key steel bars 10. Resistance to tensile stresses 11. Pockets for resistance to comprehensive stresses (Courtesy of MBM s.p.a., Milano, Italy).

course, the traditional 12 ft wide mobile home in which the highest degree of completion is achieved, despite the fact that the production process is not very mechanized. New trends in mobile home building, however, tend to make them look more like sectional homes, and, to an extent, are reducing the degree of factory finish in modules.

When boxes are stacked in high-rise configuration, the degree of factory finish is reduced for economic reasons of avoiding doubling of walls, ceilings, and floors. Another problem in stacking has also been the engineering problem of making the lower boxes stronger in order to hold the upper levels.

Although boxes in theory seem to represent the most advanced industrialized unit in systems building, totally enclosed and finished boxes are often not produced in practice. But since they derive from the original box concept, we will examine them in this chapter.

Boxes may be made to be load bearing or only support their own weight. They may be produced in monolithic form (such as concrete or fiber shell boxes) or be made in various sections joined together in the factory. They may be of heavy weight or lightweight in construction.

Medium Weight Box Systems

This category of boxes includes those built with the more conventional methods, techniques, and materials of construction in response to the modern needs of systems building and industrialization in the field of construction. There are basically two medium-weight weight materials used in medium weight box systems: steel and wood. Use of steel is only to the extent of providing a frame structure, while in the great majority of cases both the frame structure and the sheathing are in wood and plywood.

Wood frame and steel frame box systems are a predominantly North American phenomenon that is now gaining acceptance in various parts of the world, in particular in Japan, Europe, and the Middle Eastern countries, where the need for shelter is constantly increasing.

WOOD FRAME BOX SYSTEMS Two major wood frame box systems on the market are used primarily in housing: sectional and mobile. The two systems, although originally quite different in concept and purpose, are getting more and more alike in all their features, particularly in their production process.

Mobile homes are the result of the development of caravans and trailers of the 1920s and 1930s. The sectional homes, on the other hand, have been developed by panel system manufacturers who found it necessary to modify their concepts after the success of the mobile home industry.

Mobile homes are still produced on a steel frame with wheel attachments for transportation purposes, while sectional homes are transported on separate trailers and are installed permanently on their foundation walls.

MEDIUM WEIGHT BOX SYSTEM-MOBILE (C1) Mobile homes are truly industrialized modules since they are totally finished at the factory. They are built of light wood framing and aluminum siding. They come in single wide (12 ft, 14 ft or 16 ft wide) or in double wides and sometimes triple wides, which are composed of units of 12 ft or 14 ft to be assembled on site. They are not planned for permanent erection on site, but the trend is to remove the wheels and set them on temporary concrete blocks or foundations.

Studies on mobile homes have shown that although the home is designed to be movable, tenants are finding it more practical to change their home rather than transport it around.

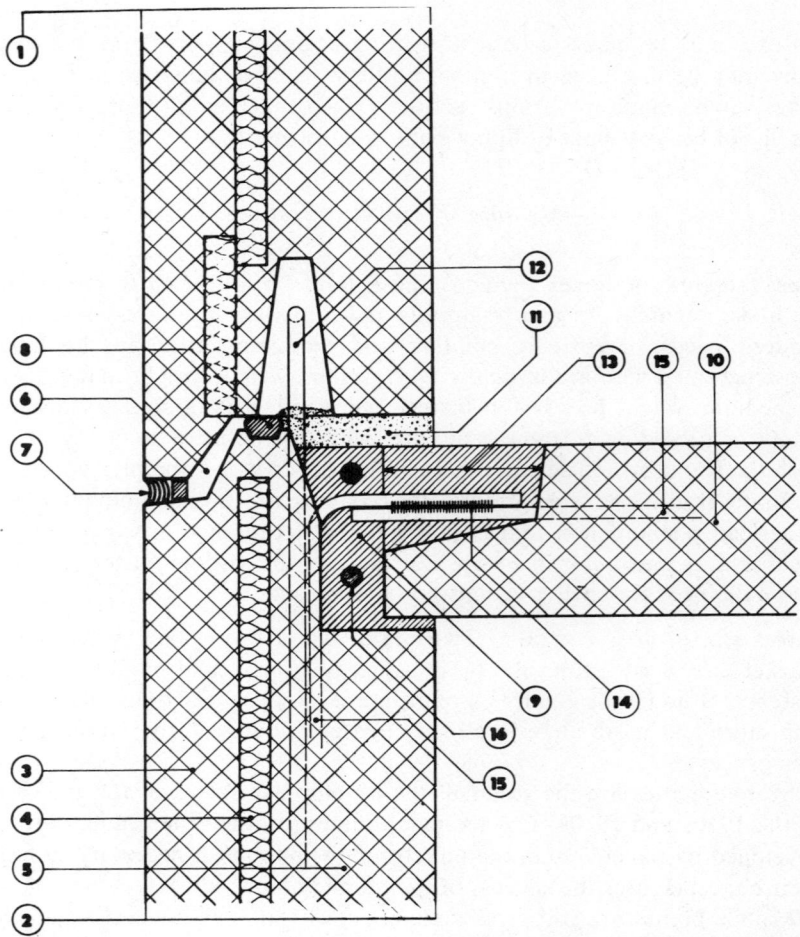

FIGURE 7.25. Exterior horizontal joint cross section. 1. Upper exterior panel 2. Lower exterior panel 3. Outer layer 4. Rigid insulation 5. Inner load-bearing layer 6. Expansion and drain chamber 7. Sealing 8. Mortar containing seal 9. Concrete filling 10. Prefab concrete slab 11. Cement Mortar filling 12. Steel bars for lifting and and connection 13. Pockets for steel connection 14. Welding 15. Steel bars connecting slab and panels 16. Continuous steel bars (Courtesy of MBM s.p.a., Milano, Italy).

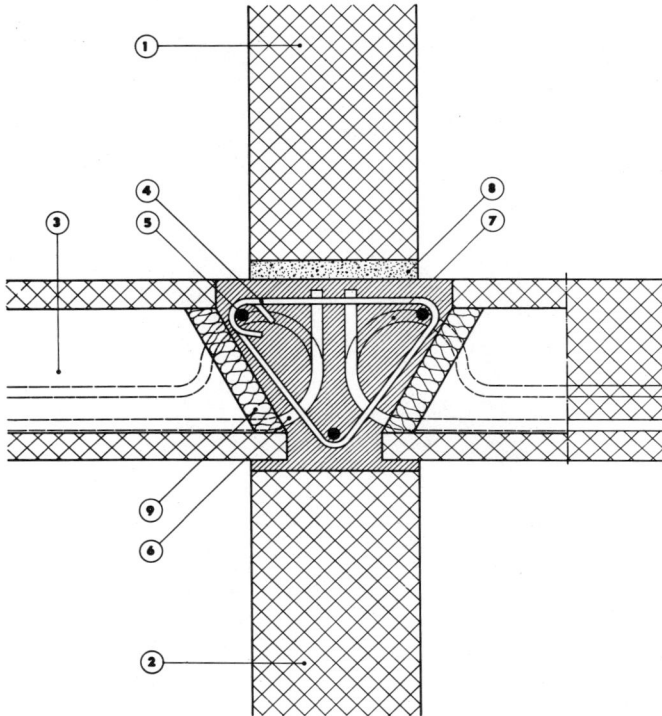

FIGURE 7.26. Interior horizontal joint with precast concrete slabs. 1. Upper interior panel 2. Lower interior panel 3. Precast hollow core concrete slabs 4-5. Ties and Reinforcing bars (continuous) 6-7. Reinforcing bars and steel lifting hook 8. Cement mortar grout 9. Polysyrene cap (Courtesy of MBM s.p.a., Milano, Italy).

This kind of consideration has given rise to the idea of double wides, triple wides, fold-ups, pull-outs, transhelters (transformable shelters), and so on, which are all rendering the mobile home rather nonmobile and are gradually changing its image.

There are other differences in the techniques and materials. Mobile homes are generally built with a lighter frame than that of sectional homes. The walls are thinner and the interior finish is in dry-joint paneling. The exterior finish is also of a lower quality sheet metal rather than the conventional siding used on sectional homes. (Figures 7.35 and 7.36.)

The mechanical systems used in mobile homes are also to a great extent standardized among the many manufacturers in North America. All ducts are passed under the floors and are attached to compact heating/air conditioning units produced for this industry.

FIGURE 7.27. Interior view of production plant for MBM, s.p.a., Milano, showing the vertical casting molds (Photo courtesy of MBM).

The major difference, though, is in the software aspect of the mobile home concept. The mobile home is treated as a car or comilar commodity, with an average life expectancy of 10 years. The financing methods for this industry are also similar to that of car financing, and this has a lot of appeal for the low-medium income family, who can move into a home with little or no down payment, and monthly installments.

The cost factor is of primary importance. Although there are small variances between the sectional and mobile home production techniques and materials, there is a large difference in price per square foot of these systems (which varies from 15 to 50% dependent on the type of home.) In 1978, for example, the cost of an average mobile home was $13.75 a square foot, furnished; the conventional single family homes were in the area of $24.75, unfurnished.

Today, mobile homes are still the cheapest manufactured system available on the market. They are not yet very nice to look at; and mobile home parks, with all their efforts, have not been able to change their drab image, but they are answering the need of a large portion of the population by keeping their costs down. See Table 7.7.

SECTIONAL (C2) Sectional homes, particularly in wood frame, are as mentioned, a North American phenomenon. They came into being for a variety of reasons—such as the need for larger control over the quality of the final product, the reduction of onsite labor and costs, and the speed of erection and installation of homes onsite in all climatical conditions.

The sectional home can be truly defined as industrialized housing. It is built almost entirely offsite; it uses assembly line production techniques in a factory that engages semiskilled and skilled labor only (Figure 7.37). Sectional homes are built in two, three, or more modules 12 to 14 ft wide and to lengths of up to 60 ft. The sections are then assembled together onsite to produce the housing units.

The highway regulations in North America restrict the volume of shipments. In general the sizes allowed are 12 ft wide, 13 ft 6 in. high, with lengths varying.

The industry of sectional homes is more fragmented than both the mobile home industry and the panel system industry. It neither possesses the marketing abilities of the mobile home nor offers the variety of plans and building types that the panel systems offer.

FIGURE 7.28. A typical production plant for production of heavy concrete panel systems and the adjacent storage yard (Courtesy MBM s.p.a., Milano, Italy).

FIGURE 7.29. Example, Jesperson system, built in Waterloo (England) by John Laing Ltd (Courtesy John Laing Ltd).

FIGURE 7.30. A major drawback on panel systems is in the necessary use of heavy handling equipment and trucks (Courtesy of Tracoba, Paris, France).

FIGURE 7.31. Twenty-story, 120 dwelling, New York: Tracoba system used under license by Module Communities, Inc. (MCI), which also got involved in Operation Breakthrough. Architect: Skidmore. (Courtesy Tracoba, Paris, France).

208 Systems Building Hardware

FIGURE 7.32. A heavy panel system being erected in Moscow, Soviet Union (W.H.O. Photo by APN).

The sectional homes once installed look and function like conventionally built homes. (Figure 7.38). The process of factory production is quite simple and mostly unskilled labor is used in the production of these units. The factory installations for this kind of process is not highly capital intensive. Boxes have also been produced in light gage metals and composite materials. Operation Breakthrough produced some interesting box systems; but because of the nonaggregated market in North America none of the systems survived. See Table 7.8.

PRODUCTION SEQUENCE   Both mobile homes and sectional homes use similar plant setups, tools, and equipment for the production of their units, with one major difference. Mobile homes are moved to the various produc-

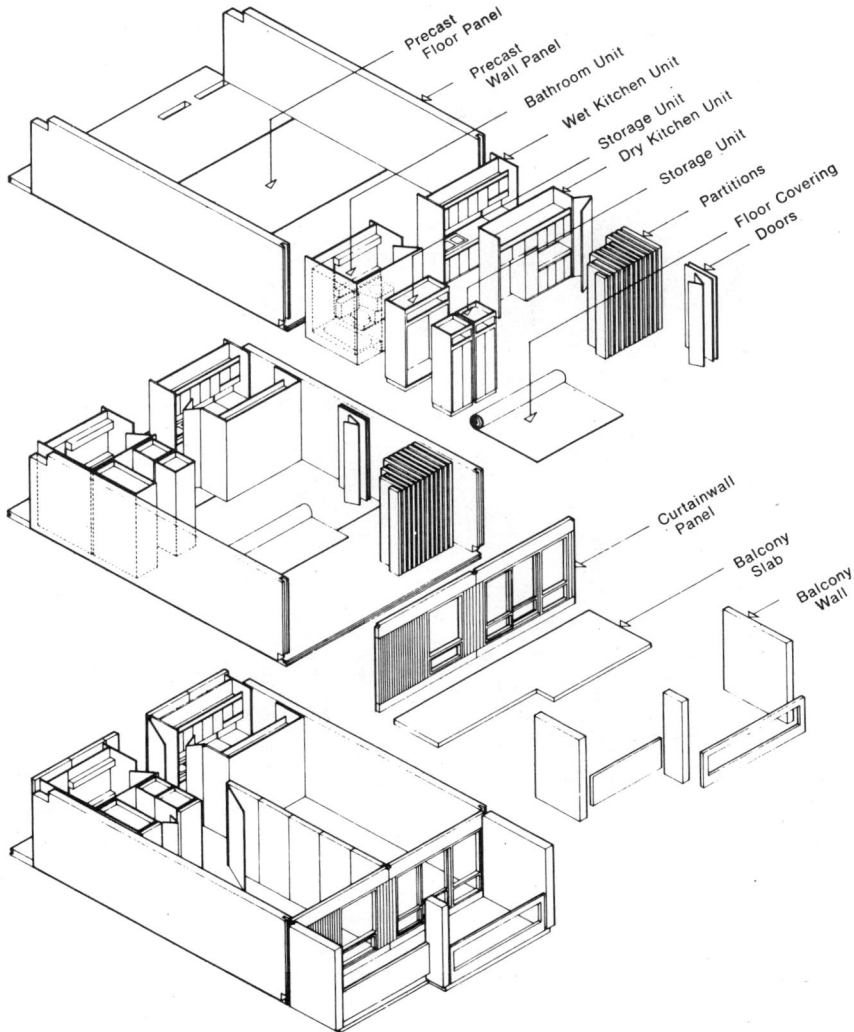

FIGURE 7.33. A totally integrated system in heavy panel construction is the Descon Concordia Building System. The components of this system are produced under strict Performance Criteria Guidelines, allowing the use of local techniques and materials (Courtesy Descon Concordia, Montreal, Canada).

FIGURE 7.34. Two applications of tilt-up construction in the field of housing. Above: Low-cost housing in Wasco, California. Below: Spanish garden apartments in Glendale Heights, Illinois (Photos courtesy of Portland Cement Association).

tion stages over their chassis and wheels, while in sectional homes, a special transport system has to be devised for moving the parts through the factory.

Plants vary in size. Most plants, particularly in sectional, have been developed out of plants for panel production and are generally too crowded.

The better-designed plant layouts cover an area 40,000 to 100,000 ft^2. Such plants are capable of producing between 600 and 1000 units of sectional homes a year on a normal shift. Production can double with double shifts. there are several stations in each plant.

Station 1—Floor assembly.
Station 2—Interior partitions.
Station 3—Exterior partitions.

Station 4—Plumbing, fixtures, electrical.
Station 5—Roof assembly.
Station 6—Interior and exterior finish.

There is a large cutting and preparation area 5000 to 6000 ft^2, and a series of areas for subassembly work connected to an inspection area.

In sectional home manufacturing, there is a need for pits at station 4 in order to install the mechanical connections.

Heavy Box Systems (Factory Produced) (C3)

This system, like the foregoing, is more widespread in North America. From the point of view of a mechanized production process it is the most advanced. But there are major limitations to its use. Transportation is the major drawback. The road limitations of 12 ft wide and 13 ft 6 in. high for maximum haul and the excessive weight of the units (70 to 100 tons) make its application undesirable.

TABLE 7.7. Performance Evaluation: Box System (Medium Weight) Mobile (C1)

	+ Advantage	=	− Disadvantage
Structure	Wood framing-steel frame; simple technology; lightweight	Fireproofing required; life span 10–12 years	No load-bearing capacity; bad sound insulation; low life span
Architectural application	Very limited; compact interiors, furnished	New approaches in double-wide fold-out, and so on	Limited on width; inflexible; mega structure required to go high rise
Industrialized process	Highest degree of factory finish; labor intensive; semiskilled unskilled labor; management oriented	Economical on average-scale production; medium capital outlay; simple tooling	Semi-industrial production process; small degree of standardization
Transportation	One or a few components	Limit of transport, 150 miles	Expensive on long haul
Erection and assembly	Simple; unskilled personnel; fast		Not foreseen for multistory

60 × 24 3 BDR 2 B Double wide

Optional angle kitchen

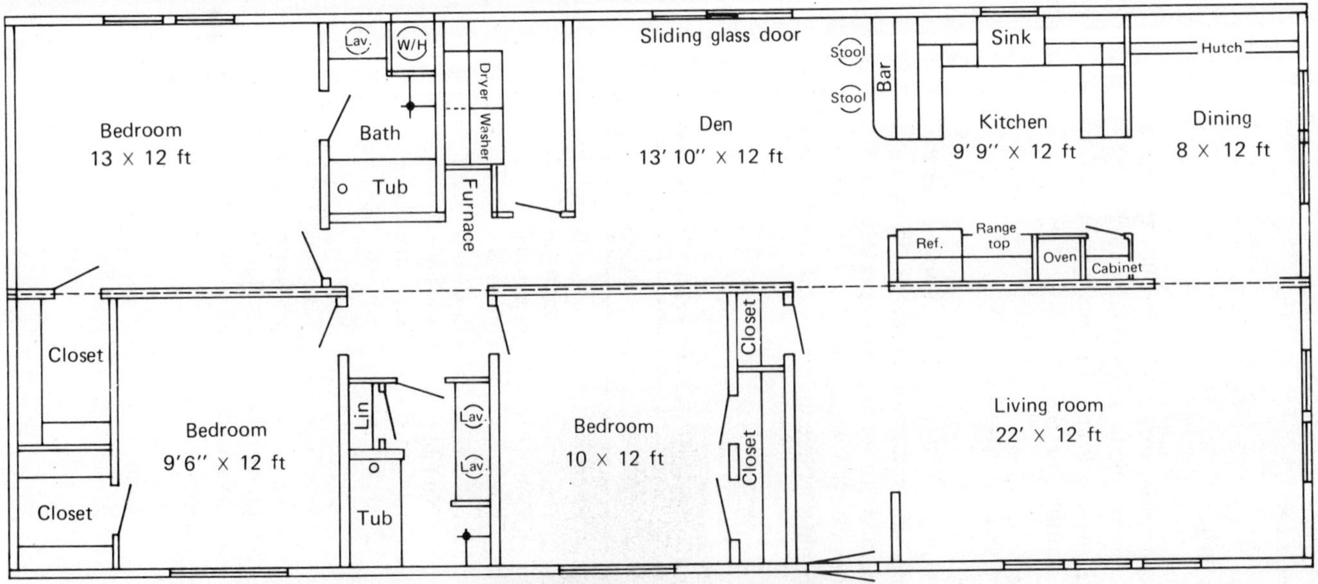

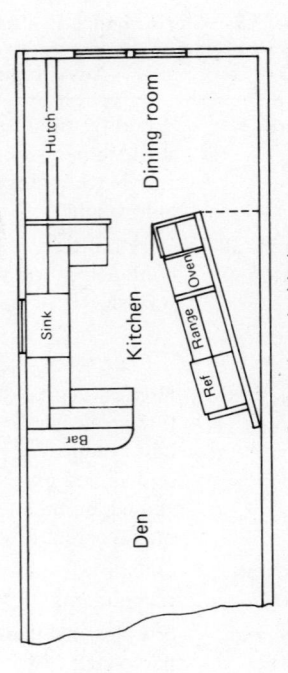

FIGURE 7.35. Perspective view and plan of a typical double-wide mobile home (Photo courtesy of Town and Country Mobile Homes, Verona, Mississippi).

FIGURE 7.35. (Continued)

Large units, for economical reasons, cannot be transported to more than a 30 to 40 mile radius. Subsequently, it is rare to find a market capable of absorbing thousands of units per year within the 60 to 80 mile diameter that can justify the initial capital outlay for a prefabrication factory in the heavy box system. (The capital outlay is in the order of $5 to 10 million for a plant producing 500 units a year.)

This point is made to show that although from the standardization point of view the box system offers the highest degree of offsite construction and industrialization other factors make its selection as the "new building block" uneconomical and unfeasible.

Another important drawback of the box system is the danger of monotony. A dramatic illustration of this danger can be found in the applications of box systems in the Union of Soviet Socialist Republics, where despite the sophisticated and advanced building techniques, stereotypes and monotonous buildings are covering the suburbs of large cities.

It is true that this risk of monotony can be broken (as in Habitat 67, discussed next), but this is done at the risk of going against the basic premise of standardization and to build each unit or set of units with some variance at a higher cost.

Moshe Safdie's Habitat 67 was a forerunner in the field of heavy box systems (see Figure 7.39). It was composed of 354 modular units that were to produce 158 units of housing with areas varying between 600 ft^2 in one-bedroom dwellings to 1700 ft^2 in four-bedroom houses.

TABLE 7.8. Performance Evaluation: Box System (Medium and Lightweight) Sectional (C2)

	+ Advantage	=	− Disadvantage
Structure	Wood or steel framing; known technology; lightweight	Fireproofing required; average sound insulation; average life span 25–30 years	Limited load-bearing capacity
Architectural application	Acceptable results in low-rise housing	Stackable (limited)	Limit in width; inflexible; mega structure required to go high
Industrialized process	High degree of factory finish; labor intensive; semiskilled and unskilled labor	Economical on average scale production; medium capital outlay; average degree of standardization (40–50%); simple tooling	Semi-industrialized production process; high degree of finish 80–90%
Transportation	Few components	Economical; limit of transport, 300 miles	Expensive on long haul; unfeasible for national market
Erection and assembly	Simple for single-story; fast	Semiskilled personnel	Expensive multistory; expensive handling equipment on multistory

Habitat was not a totally prefabricated project. Despite the fact that the boxes were designed to be load bearing, a separate heavy structure was added to provide vertical and horizontal access to the units. This structure included the mechanical space, but its true function was to provide lateral support and horizontal bracing to the assembly. See Table 7.9.

Before the Habitat

A lighter weight box system was produced by Conrad Engineers of Los Angeles and built in Richmond, California, and named Uniment System of

Industrialized Housing. Here the units were in precast, chemically stressed concrete and were transported and erected onsite to produce a 6-story, 24-unit building. All modules were finished in the factory and assembled in the field by union trades. The weight of the concrete boxes was reduced considerably (to 12 to 15 tons) by th use of Chem-Stress, which brought the thickness of the walls down to 2 in. The cost of Uniment housing at the time of completion (1967 to 1968) was $12,000 a unit. The apartment project did not use any overhangs or cantilever parts. The boxes were stacked on each other allowing for stacking of mechanical shafts. The medium height of the building and the lightweight of the modules, as can be observed, were instrumental in bringing the costs down.

Another system of stackable boxes has been used by H. B. Zachry in Texas and the southwestern United States. The same system with minor variations was used by System Building International in Canada.

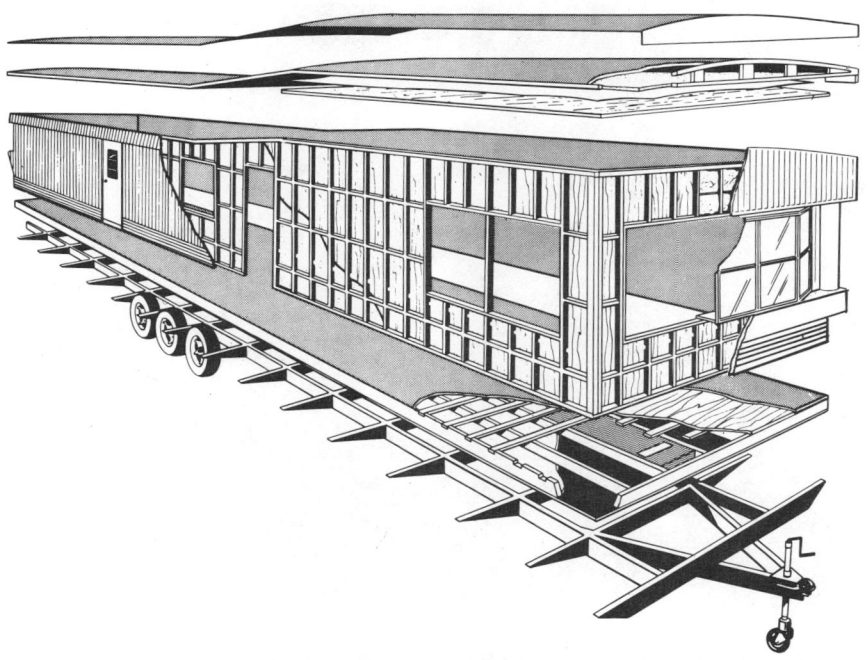

FIGURE 7.36. Schematic view of a typical mobile home. Specifications: floors—steel frame carriage, floor joists at 16 in. centers, plywood subfloor, carpeting. Exterior walls: 2 x 2 or 2 x 3 in. studs, 16 in. centers, fiberglass insulation, exterior aluminum or metal cladding, interior finish hardwood or similar paneling. Roof made of special wood trusses, paneling on the interior. One piece heavy fiberglass, neoprene or metal blanket for the exterior.

FIGURE 7.37.

FIGURE 7.37.
(Legend over)

TABLE 7.9. Performance Evaluation: Box Systems (Concrete Heavy Weight) (C3)

	+ Advantage	=	− Disadvantage
Structure	Load bearing; limited height; fire resistant; good sound insulation; simple jointing	Life span 75–100 years	Weight of components, 30–90 ton; require heavy handling equipment
Architectural application	In housing medium/high-rise building	Stackable	Inflexible; limited by width and weight; require finishing onsite
Industrialized process	High degree of factory finish; simple process; large degree of standardization, 75–80%	Semiskilled labor	Capital intensive; expensive tooling
Transportation	Few components		Costly dolley-truck; limit of transport, 30–40 miles
Erection and assembly	Relatively fast		Costly handling equipment; skilled personnel; limited height

The modules are cast in open-ended tunnel forms in the factory and are stacked on each other on site. Anchor plates are then welded together to provide for continuity in the structure. In this system, walls and floors and ceilings are doubled up, but since the box is monolithic it is built in thin walls of 3 to 4 in.

To avoid repetition of wall and floor members, a very ingenious system has been put on the market that uses the checkerboard principle of stacking as shown in figures 7.40 and 7.43. Shelley system boxes are produced in

---

FIGURE 7.37. A typical factory for the production of sectional homes is not much different from a panelized system factory. Panels are fashioned on special jigs and then assembled together to form the box. Here at Sandler Built Homes, an innovative process is used in which Upsom board panels measuring 8 ft wide (2.4 m) and up to a length of 22 ft are applied at once and clinched to the wall studs. Openings are later cut out according to need (Courtesy Sandler Built Homes, Des Moines, Iowa).

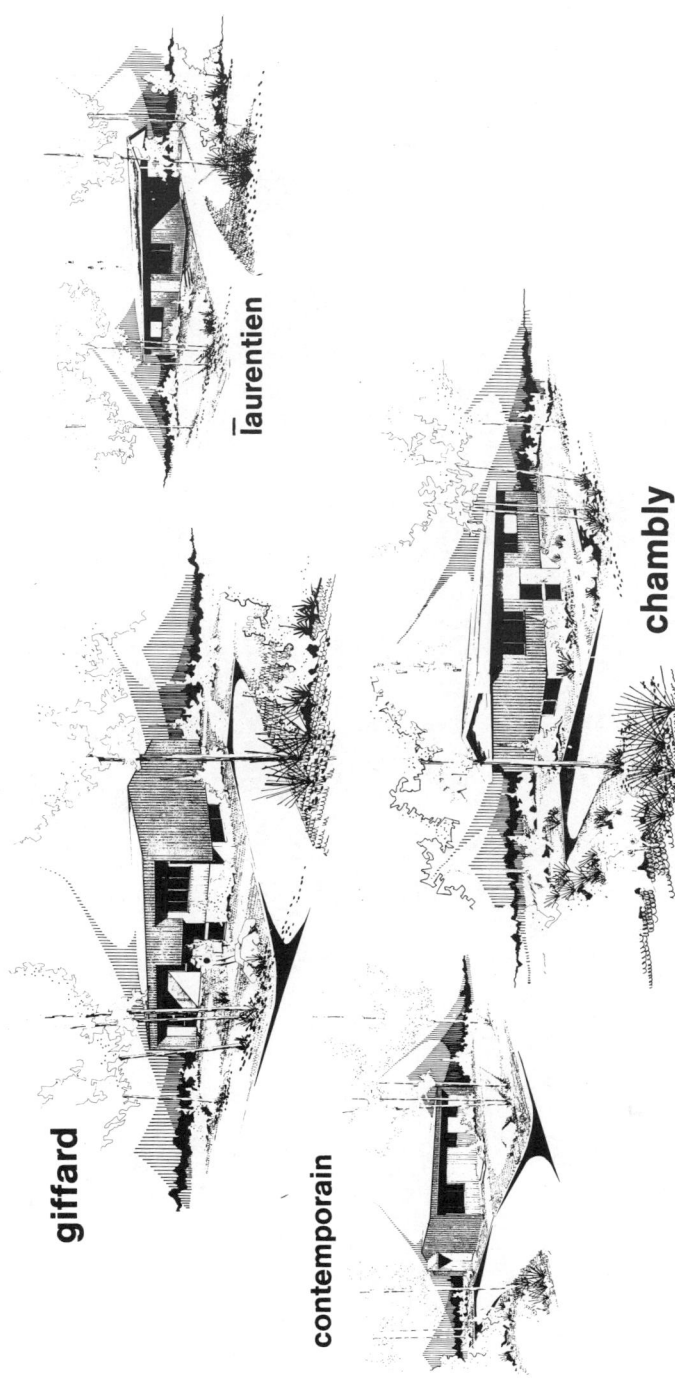

FIGURE 7.38. Examples of some sectional home designs (Variahab, Quebec, Canada)

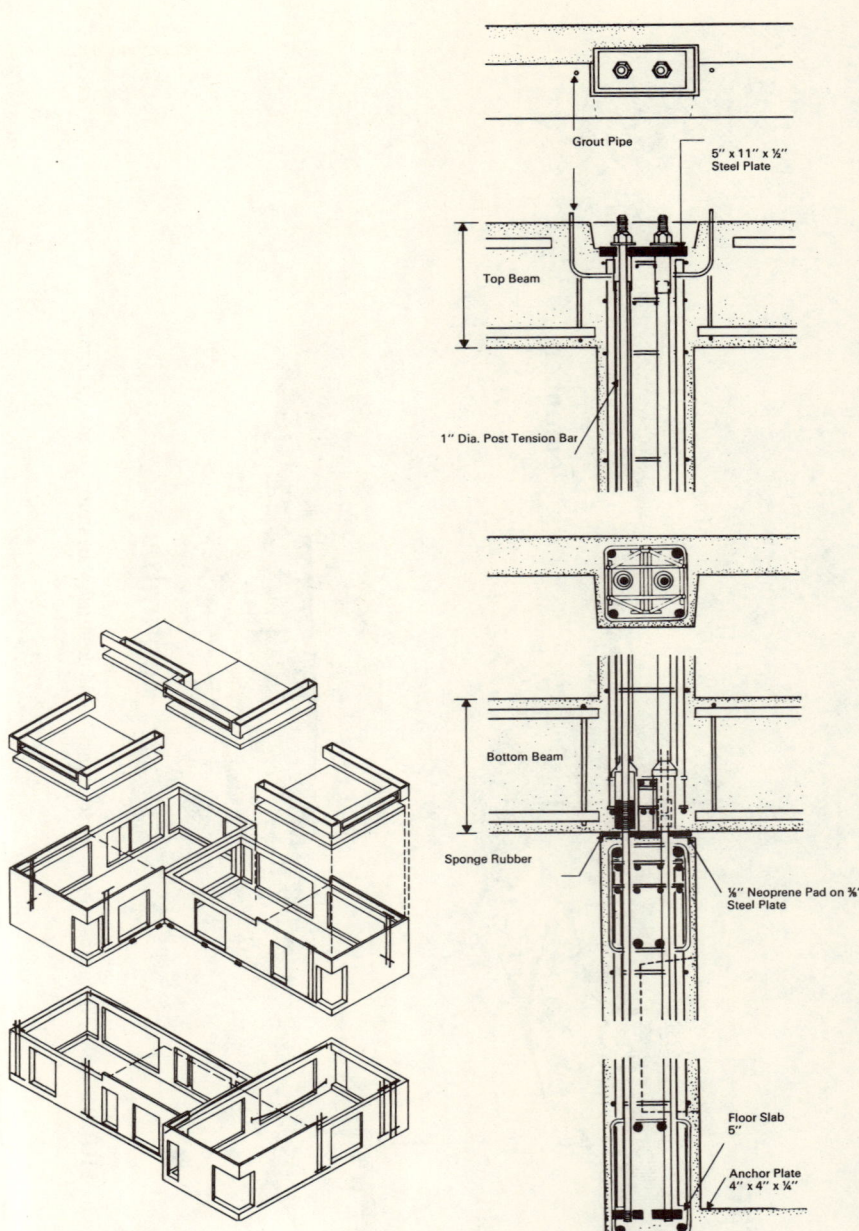

FIGURE 7.39. Habitat 67 boxes were produced open top to avoid the repetition of floors and ceilings. After stacking the leftover areas were covered through conventional construction. The boxes were joined together by post tensioning (Courtesy CMHC, Ottawa, Canada).

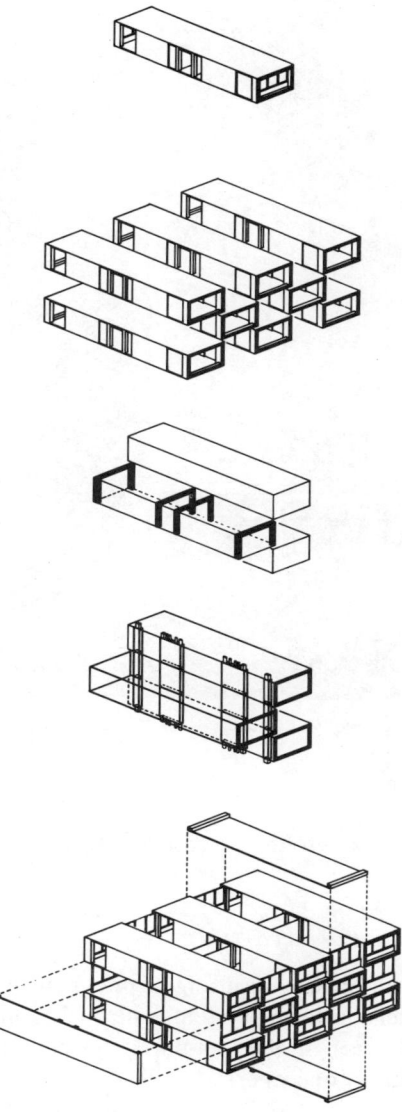

FIGURE 7.40. The basic principle of the Shelley system is in the stacking process at a checkerboard fashion, thereby the creation of additional habitable space. In this system there is no doubling of walls and floor ceilings.

large sections and use an integrated structural system within the box for their assembly, allowing large openings in the sides of the box and thereby the creation of more architecturally acceptable housing units. Although the checkerboard stacking is very economical—each unit in effect produces two units—it has the disadvantage of not being totally factory finished. A large

FIGURE 7.41. Checkerboard assembly of heavy concrete boxes produced by Shelley system, Puerto Rico (Courtesy Shelley).

portion of work must be done onsite to complete the assembly, particularly in the area of gained space.

Heavy Box Systems (Site Produced) (C4)

TUNNEL SYSTEMS   An interesting development resulting from pouring panels onsite has been the so-called tunnel shuttering system. The basic concept of this system consists of using a preengineered molding system for pouring the walls and floors of the units in one operation. The molds are highly mechanized shuttering devices that not only incorporate heating mechanisms for speedy curing of concrete, but also are arranged to be retractable once the cure is terminated. See Figure 7.44.

The many advantages to using this type of box system follow:

1. The labor content is kept to a minimum.
2. Jointing of the structure is immediately achieved through the monolithic pour.

3. The problems related to factory production and transportation are eliminated, while a high degree of mechanization is still involved in the process.
4. The system allows easy changes in size or span of floors, therefore, in part, eliminates the monotony in the system.
5. The capital investment in the system is relatively low.
6. The most important aspect of the system is its program of shuttering, curing, and release for each tunnel, which is done in a systematic manner by nine crew members and a crane operator during a daily cycle (see Figure 7.45).

Additionally, see figures 7.46 and 7.47.

FIGURE 7.42. A Shelley apartment complex being erected in New York (Courtesy Shelley).

224 *Systems Building Hardware*

FIGURE 7.43. Heavy boxes require heavy handling equipment.

PROCESS OF FABRICATION   This system uses this relatively simple process of fabrication:

1. Forms are placed in position, leveled, and connected to each other to form a set of tunnels.
2. The wall reinforcing is positioned and concrete is poured in cross walls.
3. Additional reinforcing for floors is positioned, and with the help of a molding frame, the floors are poured.
4. The whole section is then covered and steam is introduced to cure the concrete in 4 hr.
5. The molds are subsequently unlocked and removed by crane for use in another cycle.

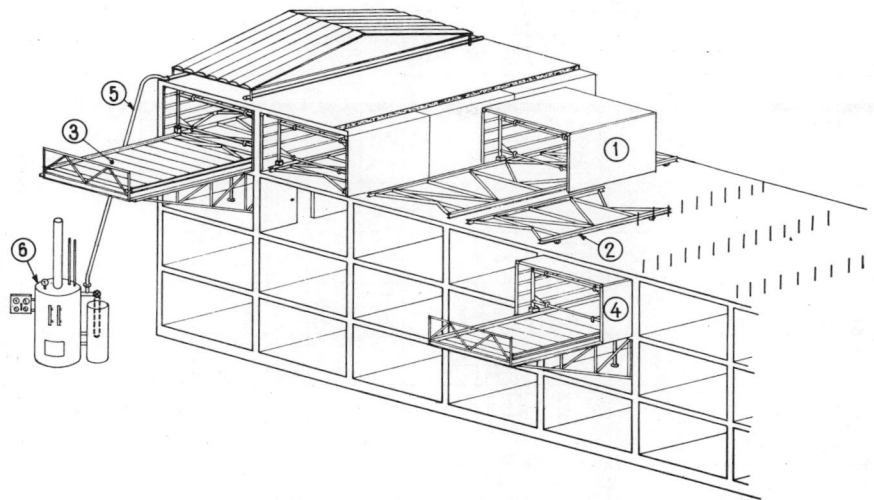

FIGURE 7.44. Schematic view of Tracoba IV Tunnel system. 1. Positioning of shutters 2. Movable track 3. Passageway for removal of shutters 4. Shutters in process of dismantling 5. Vertical heating duct 6. Heater.

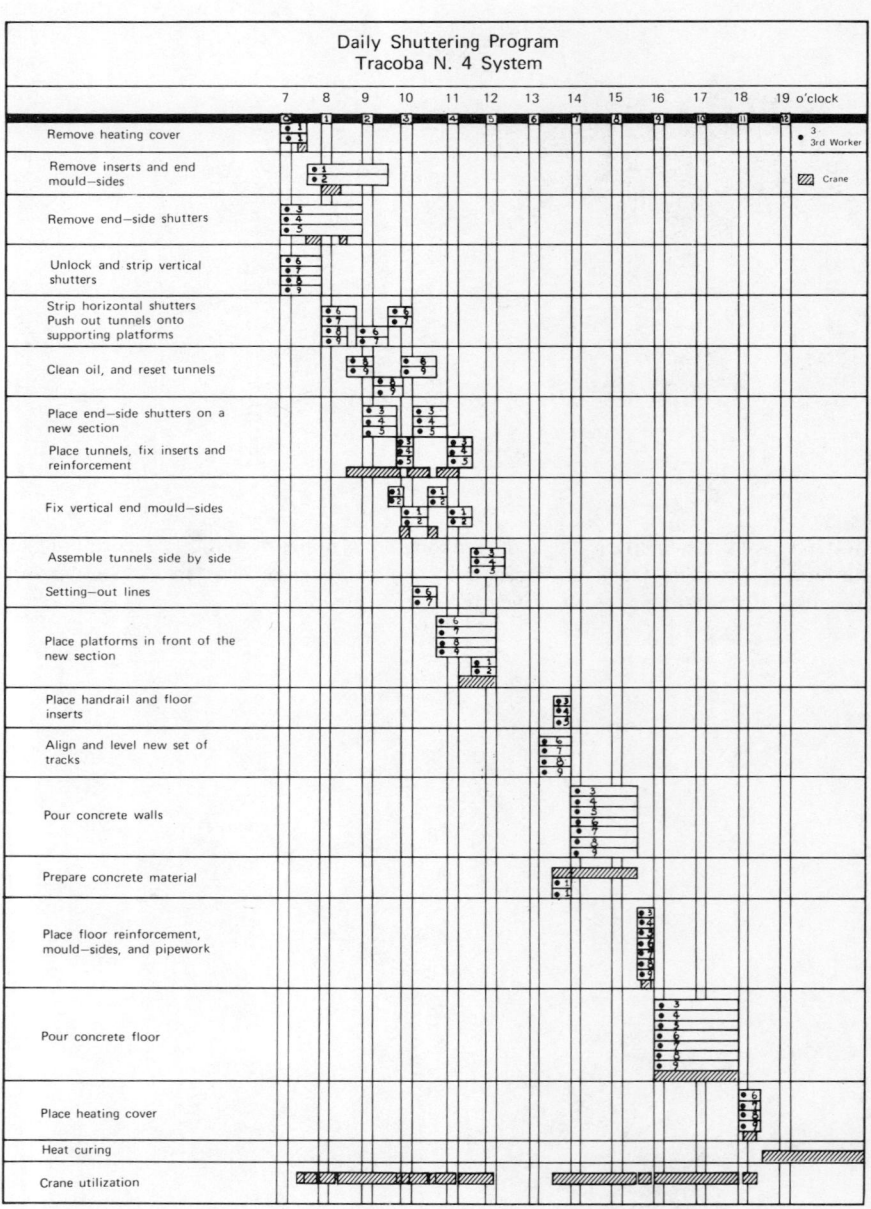

FIGURE 7.45. Daily schedule of activities for a group of nine workers and one crane operator in production of tunnel systems (Courtesy of Tracoba, Paris, France).

FIGURE 7.46. Tracoba Tunnel system, Paris, France.

FIGURE 7.47. 260 flat, 15-story building in Belfast, constructed through the use of tunnel systems (Sectra) (Photo courtesy of John Laing, Ltd).

TABLE 7.10. Performance Evaluation: Heavy Box Systems (Tunnel) (C4)

	+ Advantage	=	− Disadvantage
Structure	Durable; load bearing; monolithic	Long life span, 50–75 years	Structural support in cross walls
Architectural application	Good for self-help; medium high rise		Rigid usage of space; non-modifiable
Industrialized process	Economical on large scale; fast Relatively low capital outlay for equipment	Average degree of standardization	Proprietory system; skilled labor onsite
Transportation	No problem		
Erection and assembly	Simple systematized process		Need for crane and crane operator; skilled labor

Substantial economy in labor, transport, and time is made in these systems which brings the factory production methods to the site. The system is not dependent on proximity to site of a production plant.

It is evident that this system, although quite systematized and rapid in the production of the frame of the building, lacks the predominant characteristic of box systems, which are delivered 80 to 95% finished to the site.

The tunnel system on the other hand may be an economical method of producing a large number of basic units in a relatively short period of time where methods such as self-help are in use or in markets having an abundance of semiskilled or quasi-artisan labor for the completion of units and personalization. See Table 7.10.

8

Aided Self-Help Housing

Adel Fareed

Since the dawn of history people's need for enclosures has remained almost the same. Their two principal enclosures are clothes, which take different shapes, and houses. Although the first can be done without in some regions or under certain conditions, the second is indispensable. Houses are people's physical shelter from nature's inclemency and should be safe and comfortable. The evolution of houses reflects people's progress, civilization, and culture.

With the vast population increase, housing is now a worldwide problem that will become still more acute in the years ahead. The UN statistics showed that more than 50% of the world inhabitants are without adequate housing and living in poor conditions.

Part of this shortage, in urban communities today, arises from dependence on the housing authorities. Those people who depend on the authorities for a house must wait for a long time, with many sometimes waiting decades or more. This is due to the lack of coordination among government authorities.

According to the UN projections, the world needs to build in the next 30 years about 700 million dwelling units to house the expected population increase. Some people may think that the problem is limited to finding sources to finance these projects, but in fact the problem is deeper.

The world resources, and in particular of developing countries,—building materials, prevailing methods of construction, labor—will not be able to cope with this tremendous volume of work even if monetary liquidity were made available. Therefore, we should start immediately to look for economical alternatives to the traditional building materials and methods of construction. Also, we need design programs to train craftsmen and peasants to make use of their own personal efforts in erecting their houses by using special elements easy to produce and to assemble.

This is what is called the "aided self-help housing programs." By organizing and programming this system, one can attain an effective and economical method that will help to alleviate the housing crisis. Making use of the owners' efforts we can save part of the labor cost and accordingly get homes built at the lowest initial cost.

What is Aided Self-Help (Washington, 1973)

As the name implies, aided self-help systems depend on the work hours and labor force from which no profit is obtained in labor and peasant circles and in urban societies. The ill-housed families are trained to build, with some form of aid, better homes. The amount of training should be enough to attain such a goal, without reaching the level of professional technical labor. These trained laborers use their unused leisure time, holidays, and weekends to raise their standard of living and to help in erecting their own houses. It should be understood that this is not a cure-all to solve the world housing problems, but is a successful model that was applied in many developed as well as developing countries, where it helped to ease the housing problem.

Wages Versus Productivity of Construction Laborers

Since the end of World War II, the building industry in the whole world has undergone great development because it has had to fulfil the demand of nations—European ones in particular—to construct hundreds of cities destroyed during the war and to shelter the homeless millions.

In studying and analyzing the factors affecting the cost of the building industry over the past quarter of a century, we can see that the main reason for the continual increase in cost is the wage increase for construction labor. (Lazar, 1974) It is clear from Figure 8.1 that wages increased over tenfold while the increase in the building materials and equipment was not more than threefold. It is therefore necessary to stop this continual increase in wages by expanding the training programs and increasing the supply of technical labor through the generalized use of aided self-help building systems.

In a comparative statistical study (Washington, 1972) of wages versus productivity of American construction and industrial workers, it was noted

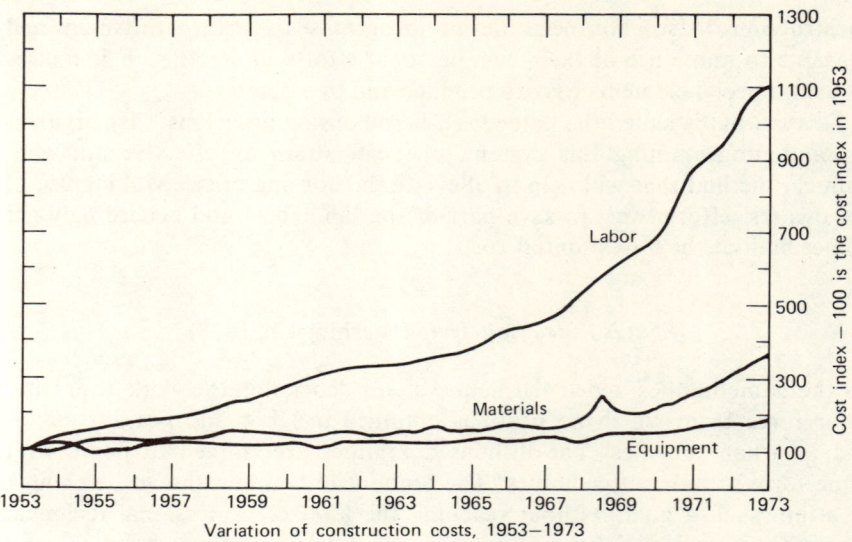

FIGURE 8.1. Variation of construction costs, 1953–1973.

that in the past 15 years, the increase in wages of construction workers was noticeably higher than that of industrial ones. However, the productivity of the construction workers had decreased as compared to that of the industrial workers, Figure 8.2. This once again emphasizes the role of training and its expansion to increase the efficiency and productivity of construction labors and end the soaring increase of housing cost.

Housing Construction by Aided Self-Help Building Systems
(Washington 1973, 1970)

The trend of self-help housing is not new. It is as old as history. Human's first priority has been to find a shelter for himself and his family to protect themselves from the cruelty of nature—from bitter cold to scathing heat, from rain and wind. The first family built its house by distributing different tasks to its members, cutting down trees and trimming and assembling them to make a house.

In rural societies this is the prevailing method for housing construction. In some, the peasant uses his spare time to make bricks from clay, which he dries and uses to construct his house by himself or with the help of his neighbors. Self-help building systems are old traditions, inherited from one generation to another. In our age, they are needed to develop the techniques and organize efficient training programs to raise the standard and life expectancy of the houses.

This principle can be implemented to a great variety of techniques and on all levels by using either modern technology or primitive methods to produce modern houses or primitive unsanitary ones. It is a way of transforming unused spare time to wealth that is added to the national income without causing economic inflation. On the other hand, large-scale aided self-help programs may help to reduce unemployment and develop building skills in countries where they are needed. The satisfaction and the pride of the owner for building his or her own house, is, of course, great and will encourage the owner's constant interest for its maintenance. This will, undoubtedly, promote enviable conversation among the surrounding

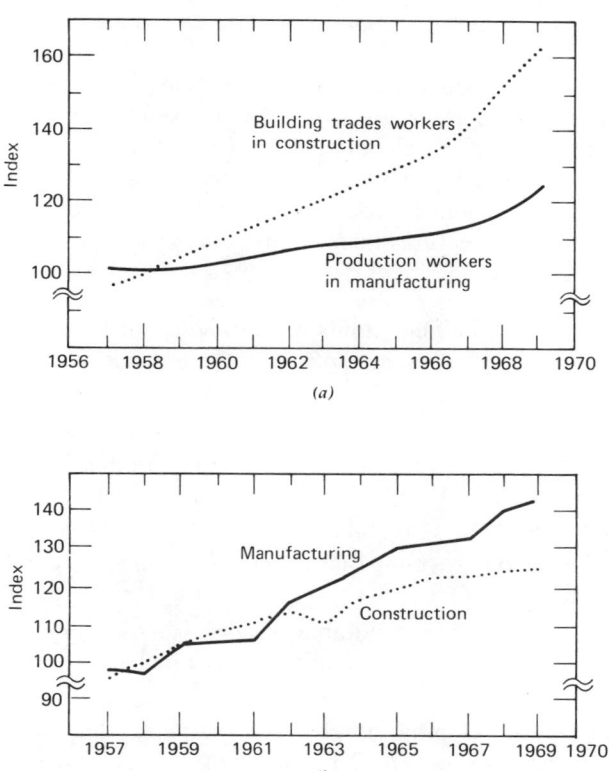

FIGURE 8.2. (*a*) United States's average of hourly earnings in manufacturing and construction—100 is the wage index in 1957. (*b*) United States's productivity in manufacturing and construction—100 is the productivity index in 1957.

neighbors and friends, following the growing awareness of the advantages achieved through the aided self-help building system.

It is the effective and practical method to encourage people with low income to own houses and thus realize their dream to be in the categories of property owners. In many of the areas of the developing world, this is the only way to have a private house. This method is different from traditional construction methods in that every region and every sector of the people need a tailored project according to the social, economic, cultural, and technical conditions. This technique depends mainly on indigenous building materials, the self-help efforts, equipment, and—the most important and urgent—the initiative and desire of the people to raise their own standard of living and welfare.

Implementation of Aided Self-Help Housing Programs

For the program of aided self-help housing to succeed, an appropriate area must be chosen where there are categories of people who are prepared to accept the idea and are ready to build their own houses. These people will be provided with management, technical help, and small loans or some materials such as permanent roofing elements and cement. The following are some of the basic features that will help the programs of aided self-help to succeed:

1. The great need of the inhabitants of the region for housing and shortage of sanitary housing for a major sector of the population.
2. Availability of people interested in the project, prepared to participate in it with their efforts and sincerely desirous of consolidating it.
3. Presence of cooperative spirit between the families to participate in collective work and their success in previous cooperative works such as environment-improving projects.
4. Existence of enough young active members among the participating families.
5. Capability for developing suitable designs for dwellings that meet the needs of typical low-income families and can be built at a cost these families can afford.
6. Capability for developing simple construction techniques that can be successfully used by unskilled people with a minimum of training and supervision.
7. Proximity of sources for building materials needed for the project.
8. Capability for designing the structural building elements to use locally available materials and choice of the size, weight, method of production

and erection of these building elements according to the technical skills and local conditions of the region.
9. Capability for establishing a working committee with trained technical personnel, sociologists, and others to carry on the program.

It is necessary to plan and organize the aided self-help project in the light of statistical and technical studies. The first stage of this study is to have a geographical and social survey to collect and analyze information and then to plan the project on the basis of the data uncovered.

The required basic information include the following:

1. The categories of people needing the project and the extent to which they are prepared to participate.
2. The distance between the project site and areas of agglomeration.
3. The social, economical, and health situation of each family. Necessary in relation to this is the number of family members and how many are capable of working, their technical experience, their house rents, the extent of their financial means to pay instalments for a new house.
4. Survey of building materials and technical labor available in the area.
5. Survey and study of the traditional building methods (especially for low-income people) of the region.

Using this information, the project is designed (the house and the construction method) according to the availability of building materials and technical skills. Dependence on paid technical labor should be limited as much as possible.

Before starting the implementation of the houses, the infrastructural facilities, which represent the minimum requirements for the creation of an integrated human settlement, should be provided by the authorities. The families are then required to construct their houses according to the designs and techniques previously prepared by the supervising organization. It is also necessary to consider places for a public market, schools, recreation areas, and playgrounds for the children. Unless self-help housing programs are planned, organized, and provided for beforehand by the necessary infrastructures, the physical conditions of the project is expected to deteriorate and will accordingly raise the world's slum stock.

Training Programs

The next stage of the project is that of training people in order to direct their efforts in building the houses. The difficulty of this stage lies in designing intelligent training programs that can realize the desired benefits in the shortest possible time. This depends on psychological, social, cultural, and traditional factors that change from place to place. The instructors should have the necessary qualifications to face these aspects.

Before starting to implement the project, it is always preferable to build a pilot demonstration house to show exactly what the people will get for their money and to gain their confidence. This pilot house is also important because it enables those involved to become aware of all the problems and impediments in implementation and to make a precise study of the stages of execution and of the real cost, as well as the effectiveness of the training program. Thus the economic volume of a project, the choice of people to work on a project, the modification of the training program and the precise estimation of its size can be determined. At the same time, financing in the form of aid from foreign countries, loans, or the like, can be provided.

The remaining stage—which also accounts for the basic reason many such projects fail—involves dealing with the wishes and requirements of the individuals for whom the projects are designed and their acceptance of the idea of participating in them. The situation that arises in regard to the owner is quite evident in developing societies with very small incomes, where suspicion is felt against any person coming to aid them. The first questions the members of these societies ask are "Why does he come to help us?" "What will he gain from this?" "Is he telling us the real objective behind his helping us?" And the list goes on. This is due to the culture of these people who are not accustomed to receive aid.

Many categories of people in developing countries who live in poverty are unable to express their own needs in housing. It is necessary for sociologists, working in a given project, to help such groups define their real needs and problems. This is achieved by direct dialog with these people and by field studies of their living conditions, habits, and traditions. Relatedly, the new property-owners need to be convinced of the feasibility of these projects so that they may respond and show a real desire to implement projects to raise the social standards and improve the living conditions of people in developing countries. Without such attempts at working together, the desired success of the programs will not be achieved.

Further, it is necessary to convince those who will carry out a self-help project—political leaders, sociologists, preachers in churches and mosques—of its feasibility and benefits. Information organizations and the press can also participate in exposing and explaining the idea and its advantages.

Television also has an active role to play by showing live models of housing experiments in aided self-help programs. It is possible to show a model furnished house with practical, nonexpensive furniture with a model family living in the house. The audiovisual means can also give a clear picture of the advantages of this method by comparing the living conditions of families before and after implementing the experiment. It can also play a great role in the training programs.

Land

Land is usually the key element in most housing projects. Land problems differ from one region to another and from one society to another. They are different in the town and in the village. Much depends on the political and social systems of the state. Aided self-help housing is sensitive to land cost and availability.

In some experiments for limiting the continuous raising of land prices, land was given to families with the condition that if they did not build on the land within a limited period (five years, for example), the state would retrieve it. In other cases, land was given to cooperative self-help building societies with the condition that ownership of the land was retained by the societies and that members could neither buy or yield it. If the member did not build on the land within the limited period, the land was restituted to the respective society to be used by another member. In some societies the state keeps its title of ownership of the land, but leases it to new owners with a symbolic rent, or none at all, for a period between 25 and 99 years. In this case the member lacks the right to dispose of or yield the land.

The basic motive for these restrictions is to limit the rise in the price of building grounds, especially in cities, and to stop the exploitation of the facilities offered to these programs and the deviation from its main objectives.

Group and Individual Self-Help

Once the people are convinced that the approach of self-help is useful for raising their living standards, they have to discuss whether the work will be done on an individual or group self-help basis. That is, an individual and his or her family will put out the time for building their own house or a group of people will build all of their individual houses by working together as a team.

The individual needs to work more man-hours to build and reach the same cost of a similar house built by group self-helpers. It is estimated that an individual needs to work 2000 man-hours to finish a wood frame house for $8000, while group self-helpers need only 1050 man-hours to build the same house for the same price.

Although group action is often more effective, it may sometimes spoil the whole program and result in total failure. This depends mainly on the nature of the people, the construction system, and the amount of time the individual self-helpers intend to contribute. Group self-help programs work best in closely knit, undisturbed societies where the family and community cooperate in providing a better standard of living.

The Puerto Rican Self-Help housing program (Washington, 1974) is a good example of an effective group activity. On the other hand, in 1950, individual families built in Greece about 70,000 housing units through a successful individual aided self-help housing program.

Financing and Foreign Aids

Although the human effort offered by families in aided self-help projects represents a large portion of the construction cost, the financing for building materials and equipment still represents a heavy financial burden. This could be solved in the form of governmental loans or other loans from the savings and social services funds of the trade unions. These loans should be given without interest or with lower interest than that on the market for similar loans.

It should also be possible to offer to the families loans in the form of some building materials, exempted from customs. Heavier construction machinery need to be rented to the families for a minimum amount, representing the wear-and-tear expenses of the equipment.

It is recommended that the supervising housing organization provide the technical assistance and supervision free of charge. By doing this the organization gets to participate in the project. This also means that the families carrying out the project will be exempted from paying wages or remunerating the technicians and engineers supervising the work. The participation of the supervisors is valuable and very important for the guidance they give in the use of building materials without waste. They also try to find alternative materials available in the region and modify or develop the model or the method of construction from one site to another, according to the conditions and skills available in the area.

Participation of Families in Construction

The participation of families in aided self-help programs can take many forms, depending on the method of construction, the building materials, the skills available, and so on. In societies where building methods depend on cutting down trees and bamboo, youths are trained to do these activities (cutting and pruning wood and preparing it to be used for construction) as part of their tradition. Here the participation of the family is bigger. Perhaps the role of training, in this case, is to improve the standard of the house. Aid could be given in the form of equipment and machinery, important aids for higher productivity and better quality.

Nevertheless, urban housing in some areas depend on a level of specialized expertise that the family working in the project cannot attain, even after training. In such cases, the role of the individual becomes limited to nontechnical operations, such as transporting and preparing materials, operating equipment, and helping the technical labor workers.

For this reason, the design of the construction method is very important. It is necessary to make the maximum use of most of the people and the expertise available after suitable training. Technical and economical success of these methods depend on the design of the construction system and the training program. The production of elements for houses should depend on locally available materials at economical prices. The elements need to be small in size and weigh within a capacity that one or two workers can carry them without the use of a crane.

Construction Method

Successful aided self-help programs depend highly on the choice of the construction method. The degree of prefabrication of the building elements is determined according to the amount of working hours the self-helper is capable of devoting to build his or her house.

In rural areas where the families can give a great part of their time, a minimum degree of prefabrication may be feasible and economical. Rammed earth and mud brick enclosure technology is the cheapest when the self-help builder is ready to contribute the labor necessary to make the bricks and build the house. The builder, in this case, needs to work 200 to 300 man-hours/m^2 of the built-up area of the house. (Washington, 1974)

From 100 to 200 man-hours/m^2 of self-help input, in place of fabricated methods, may be competitive. The building materials as well as the construction technique depend mainly on the local conditions and the available skills in each area.

In developed communities, where the self-help builder is unable to give much of his or her time, less labor-intensive methods are used. In such industrial societies partially prefabricated elements are available that can be handled easily by the self-help builders. These elements are usually in the form of semifinished wall panels. Service core systems are generally used for early occupancy of the house.

A self-help builder working only 3 to 4 man-hours/m² can reach the maximum reduction in the construction cost by using a highly prefabricated technology such as prefinished wall panel elements with a service core. In this case the builder makes substantial savings by purchasing the materials and components directly from the distributor and thereby saving the percentage the contractor would charge. However, the builder's savings on the labor output he will replace are very little. In this case the builders should be well trained and careful in handling the panels. The damage of such panels may not be easy to repair on the building sites by simple methods and may require replacement. This could affect the quality and economic benefits of the project. It is recommended that these highly prefabricated methods be used only in industrialized countries where the mechanized concept is prevalent.

Original Approach to Low-Cost Rural Housing

Early in the 1930s, the well-known Egyptian architect Hassan Fathy studied the Egyptian heritage in building houses with clay. He noticed that the descendants of the ancient Egyptians in Upper-Egypt preserved the traditional architecture of their grandfathers (more than 3400 years ago) and succeeded in building their houses, mosques, and churches from mud bricks. These bricks were used to build walls and roofs, and no cement, steel, or shutterings of any kind were used. The roofs were shaped in the form of vaults and domes whose combinations were in harmony and imposed a beautiful appearance. These shell-shaped roofs were built to stand up during construction without using any support, Figure 8.3. The people of the regions would decorate the houses, covering almost every corner of the interior with their primitive arts. Walls were built with slots and air holes that, together with the insulating characteristic of the mud, ventilated and cooled the inside atmosphere of the houses and rendered them pleasant to stay in during the hottest days of the Upper-Egypt summer seasons. Hassan Fathy (1973) adapted and evolved this concept of construction for rural mass housing. His experiment was pioneering in building some of the villages and houses by applying a self-help program. He proposed a work pattern in which the peasants essentially provided their labor to fabricate mud bricks from the earth at nearly no cost and then used this building material to construct their houses by themselves. For training the people, he designed a five-stage short training program and applied it for the first time when he was preparing the necessary masons to build two villages. The

FIGURE 8.3. Buildings of mud brick, vault-shaped.

houses of these villages as well as all the public buildings (schools, theaters, mosques, and the village market) were constructed of mud brick walls and roofed by elegant mud vaults and domes, Figures 8.4 and 8.5.

The experience gained from this experiment enforced the belief that the maturing of skill is an experience of considerable spiritual value to the craftsman. A person who acquires the solid mastery of any skill grows also in self-respect and moral stature. The experiment proved that the transformation brought about in the personalities of the peasants when they built their own villages is of value to them as the transformation in their material condition.

The results of the experiment were tremendous with regard to the engineering and social points of view. They became a subject for study by UN housing agencies and other international building research institutes.

This concept was first applied with mud bricks. Yet it could be successfully used with limestone blocks, Figure 8.6, or any other kind of blocks or bricks.

Successful Examples of Aided Self-Help Programs Around the World

A few of the successful aided self-help housing programs are briefly reviewed here. (Washington, 1969), (Tan, 1977)

242   *Aided Self-Help Housing*

FIGURE 8.4.   School roofed by vault-shaped mud brick.

1. *Puerto Rican Experiment.* Puerto Rico's program for aided self-help housing is one of the most successful around the world. It began in 1949 and succeeded in helping 37,000 families to build their houses. The success of the experiment was primarily a result of the government's policy and the foreign aids and support.

2. *The Greek Experiment.* By the end of 1949, and after the destruction caused by World War II, Greece needed a huge housing project to accom-

FIGURE 8.5.   Houses roofed by dome-shaped mud brick.

FIGURE 8.6. Houses roofed by vault- and dome-shaped limestone.

modate the refugees. The shortage of raw materials made it necessary to build small houses, and the lack of sufficient technicians allowed the aided self-help programs to take over. After receiving adequate training, many families worked under the supervision of qualified technical bodies. The assistance given to each house was estimated at about $600.00, which represents half of the estimated cost. The houses built this way saved 50% of the cost of those built by the conventional methods.

3. *The Moroccan Experiment.* It began in the early 1960s in the city of Marrakesh. Between 1962 and 1965, 1000 housing units were built through the aided self-help program organized and financed by the Moroccan government. They gave each family a loan of $700.00 to be repaid in instalments at an interest rate of 6%. The government provided each house with water and electricity.

4. *The Swedish Experiment.* Sweden was one of the ancient European countries to use the aided self-help systems. In 1927 the municipality of Stockholm established the first housing cooperative association. With the cooperation of neighboring families, the average family was able to build his or her house of built-up area from 60 to 80 m². The owners were able to save 30% of the cost of their houses.

5. *The German Experiments* After World War II, when Germany was faced with the destruction of the war and thousands of homeless people, the country's entire population concentrated their efforts to reconstruct their homeland. Among the many ways used was the aided self-help building

system. Helmut Bruchmann, the director of the program for aided self-help housing, said that he believes anyone who has not had himself this experience of building his house with his own hands, sacrificing his leisure and his resting time, cannot appreciate the extent of the effort and sweat needed to do it.

All the family members were sharing in the project. They started work at 4 A.M. every day. The children worked until it was time for school, the men until they left for their jobs, and the women usually worked all day long. Men and children returned later and everybody worked.

Housing was not the only problem for the Germans at that time; food also was a problem. Thus the state's policy was to encourage families to cultivate limited areas of land as backyards that would provide them with the vegetables and fruits they needed.

Conclusions

Self-help housing is an old principle well known to those in most rural areas around the world as a part of their inherited tradition. And aided self-help housing is a reaffirmation of the old-fashioned self-help, but with better planning, organization, guidance, and aid. It uses the potential labor force, skills, and unused leisure time of ill-housed families to improve their living conditions by helping them in the construction of their own houses.

Aided self-help capitalizes the unused leisure time in the form of a wealth to be added to the national income without encouraging economic inflation. The aided self-help principle leads to an increased access to home ownerships. And although the economic benefits are great, the citizenry's social transformation, self-respect, self-confidence, and dignified pride of ownership—all from building one's own house—are of far greater importance.

Aside from improving housing quality, the aided self-help principle encourages the tenants to be devoted to the maintenance of their houses. It should be understood that aided-self help is not a cure-all to solve the world's housing problem, but a great tool to alleviate the housing shortage in many places.

The aided self-help principle depends mostly on the initiative and human efforts of the owners; it is not a technique that may be used by anybody and everywhere. Consequently, it is not an easy principle to introduce in the absence of careful planning, organization, and study by sociologists, planners, architects, and engineers who want to execute it in a given area.

9

Housing Systems in India

Hampapur Sreenath

India's successive five-year plans have allocated 25% of the total plan outlay for the construction of buildings, including residential houses. At present, the problem encountered mainly is the shortage of important constituent materials of reinforced cement concrete such as cement and steel. Even though the production of cement is stepped up to the anticipated quantity of 22 million tons at the end of the plan period against the present rate of 18 million tons/annum, we still fall short of 4 million tons. The phase difference of demand and production is bound to increase in the future with the increase in mass housing activity.

This fact, together with the expensiveness of materials like steel, forces one to search for alternative construction systems in which less materials are utilized.

Several systems of roofing/flooring and walling for prefabricated houses have been developed at the Structural Engineering Research (Regional) Centre (SERC), Madras, India, incorporating the suitable locally available materials such as clay blocks and use of aerated cellular concrete products and of fly ash in concrete, and so on, to effect economical use of expensive materials. Different housing systems have also been developed by other agencies in India. The different systems in use are here explained in brief, based on a report (Zacharia, 1974) prepared by SERC under the auspices of United Nations Development Program.

Ucopan System

This modular housing system uses prefabricated concrete panels for roofs, walls, and floors. It allows for architectural freedom and is suitable for any size of single or multistoried buildings. There are only two types of panels: one for walls and the other for roofs. Depending on the loading, the reinforcing steel only need be changed in the panels. The system was evolved by the Calcutta Metropolitan Planning Organization (CMPO), under the guidance of Dr. Z. A. Zielinski (former consultant, Ford Foundation with CMPO, Calcutta).

This system has potential applications for single, multistory housing schemes (limited to 8 stories), schools, health centers and university buildings. The smallest size of house units, consisting of two rooms with toilet and bath with an approximate plinth area of 20 m^2, has been demonstrated by CMPO under the name "Tapsia Type."

Some states have started adopting this system for low-cost housing. An example is the adaptation for the construction of 112 single-storied, single-roomed tenements and 38 double-roomed tenements for the Tamil Nadu Housing Board. The system is shown schematically in Figure 9-1.

Integral House

Originally evolved in 1969 by Shri S. Krishna Iyer, Madras, this system uses medium-sized hollow concrete wall panels and funicular shells for roofs/floors. The hollow panels were 7 in. thick with 4 in. clear cavity. These panels were placed side by side leaving a definite gap, which was later filled in with concrete and reinforcing steel. The panels are connected to the foundation raft. The roof/floor is made of 3 x 3 ft precast funicular shells arranged to form a grid and the grid ribs are cast in place after placing reinforcement. Figure 9.2 shows the system as applied to a room.

The main disadvantage of the system was that the units weighted about 200 kg and needed a derrick post for erecting the units one above the other. Moreover, panels would dislodge when they were erected over mortar.

The system has now been modified to overcome these problems. The modified system uses wall panels of 1 x 1 ft and roof panels of 10 x 10 in. Figure 9.3 shows the details of the typical units.

The system in its original form was adopted by Tamil Nadu Slum Clearance Board for constructing 150 buildings in Madras City.

Castone System

This system, promoted by M/S Bombay Chemicals Precast Concrete Division, Bombay, adopts a batten and filler-block arrangement for roof/floor and precast wall panels. The wall panels are connected to each other by bolted connections. The roof/floor scheme of this system is independent

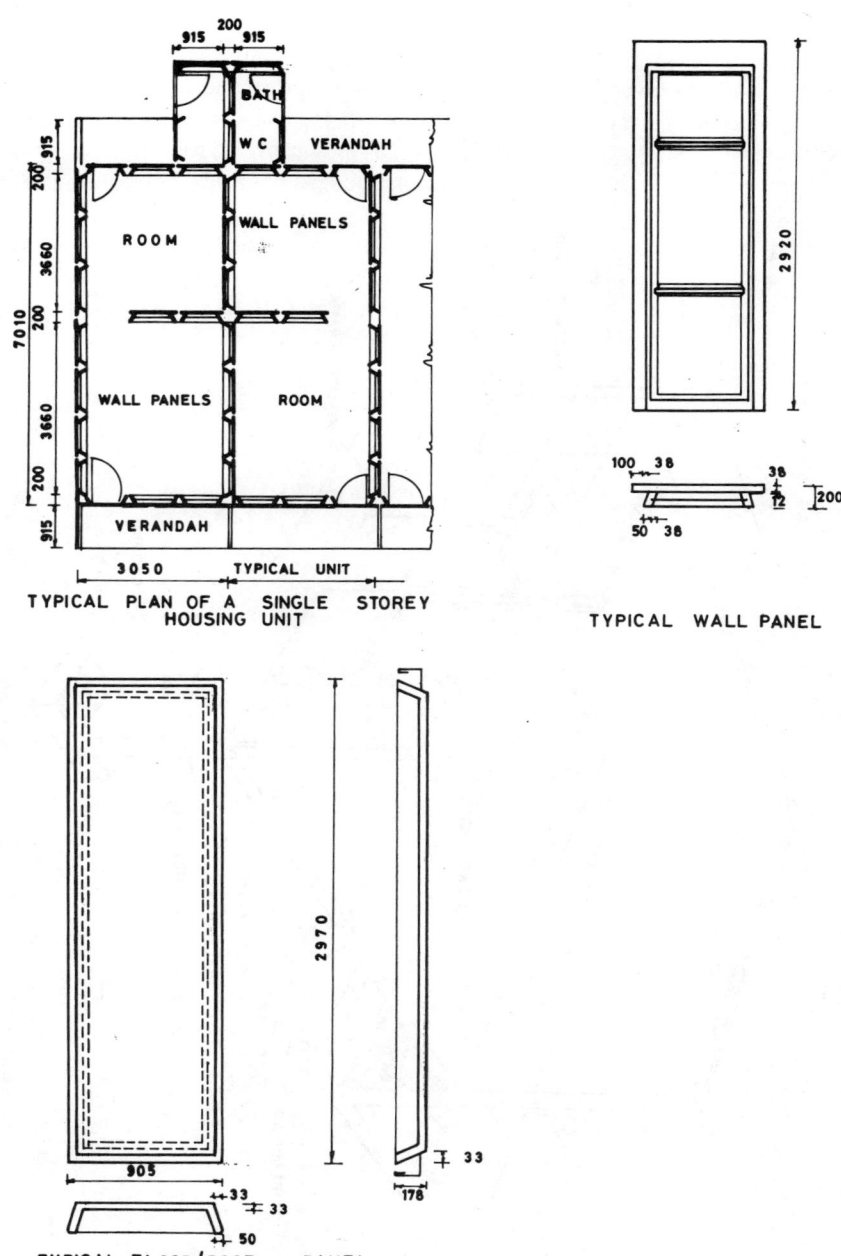

FIGURE 9.1. Ucopan system. 1. Typical plan of a single-story housing unit 2. Typical wall panel 3. Typical floor/roof panel.

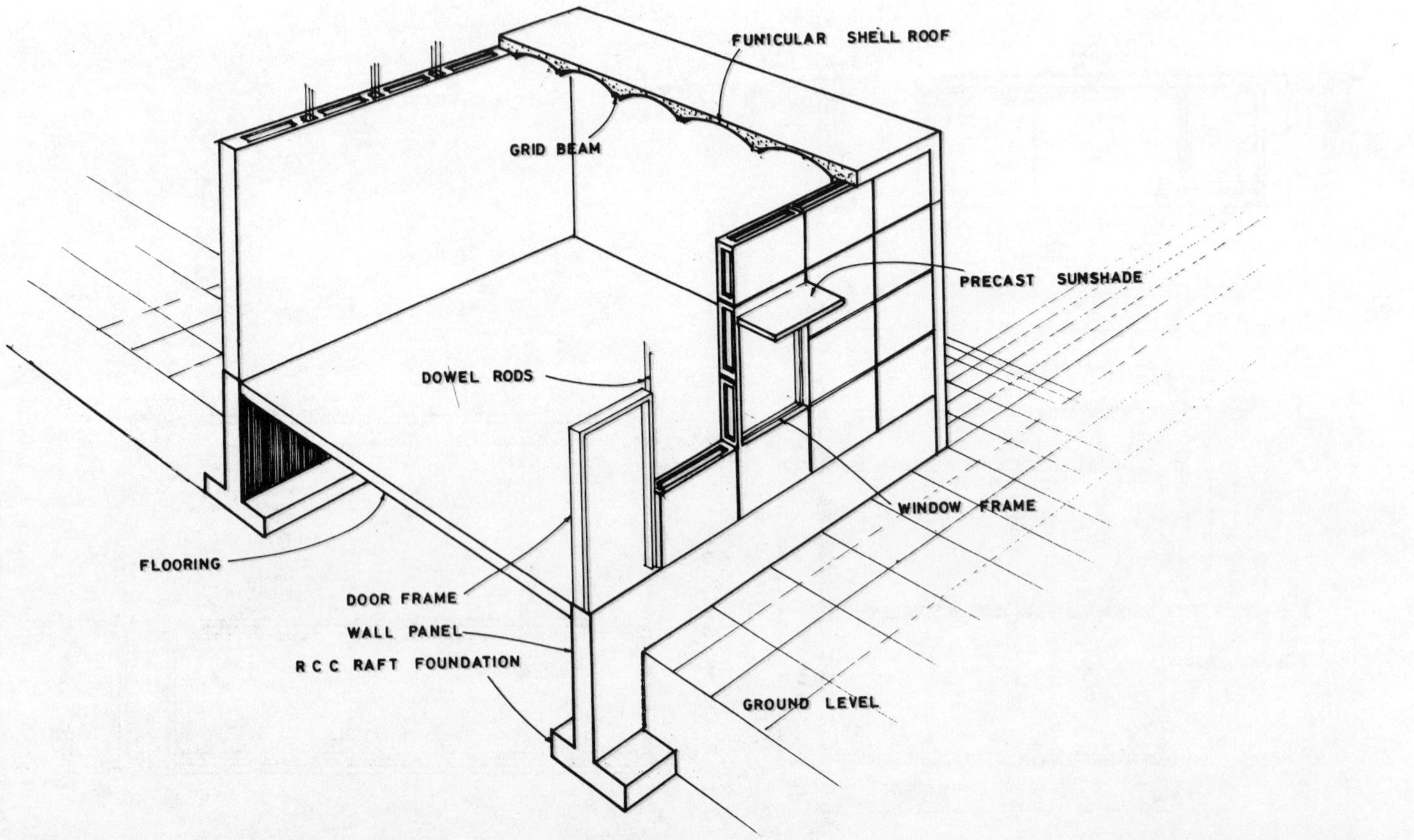

FIGURE 9.2. General view of a room, integral house.

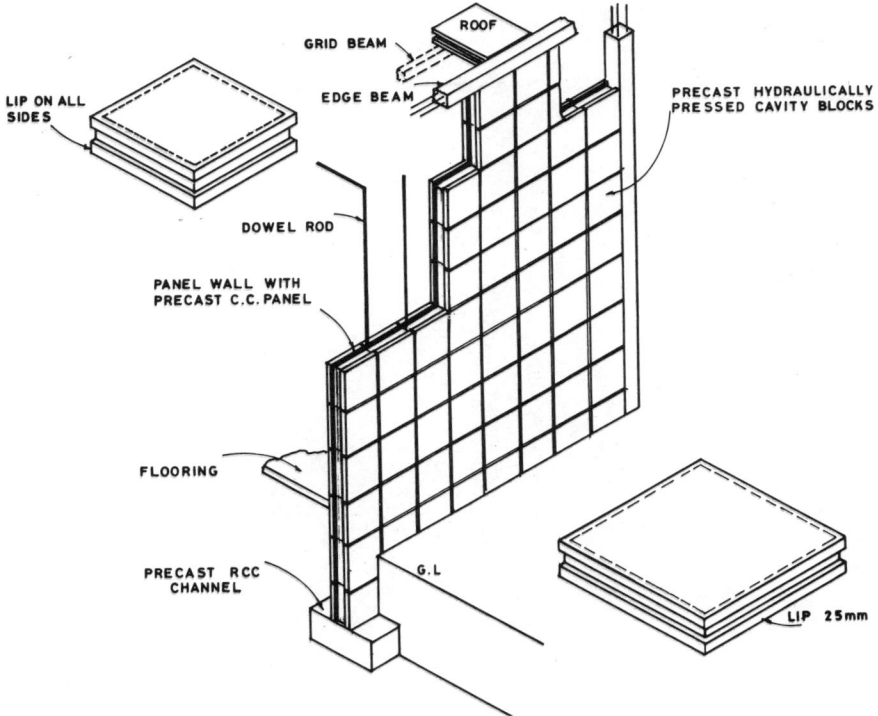

FIGURE 9.3. Built-up wall and floor/roof elements, integral house (modified).

and could be advantageously used either over load-bearing conventional walls or over the beams of building frames. Figure 9.4 shows the system schematically. The roof/floor system could be applied for multistoried residential, office, institutional, and other public buildings. Nearly 1200 residential houses have been built in Bombay adopting this system.

The Castone system uses a steel skeleton encased in concrete as a batten to support blocks, while another parallel system uses precast concrete battens and concrete hollow blocks. The depth of the plank or batten and also the block vary depending on the span.

The system, using precast battens and concrete hollow clay blocks for roof/floor, has been adopted by M/S Shalimar Tar Products and Hindustan Housing factory for tenements in Delhi, Ranchi, and Assam.

Super Prefab System

The promoters of this system, which has been used for constructing tenements for engineers, are M/S Super Prefab System (P) Ltd., Bombay. The

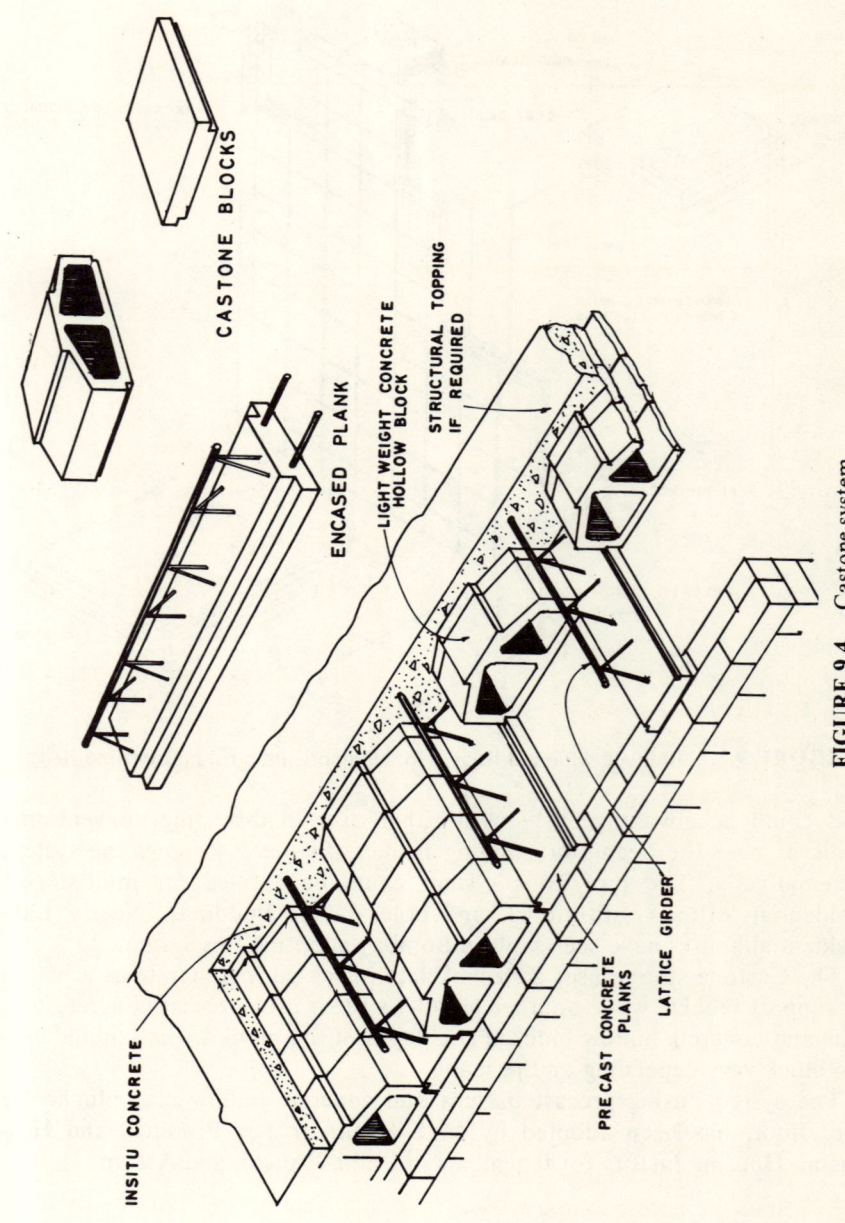

FIGURE 9.4. Castone system.

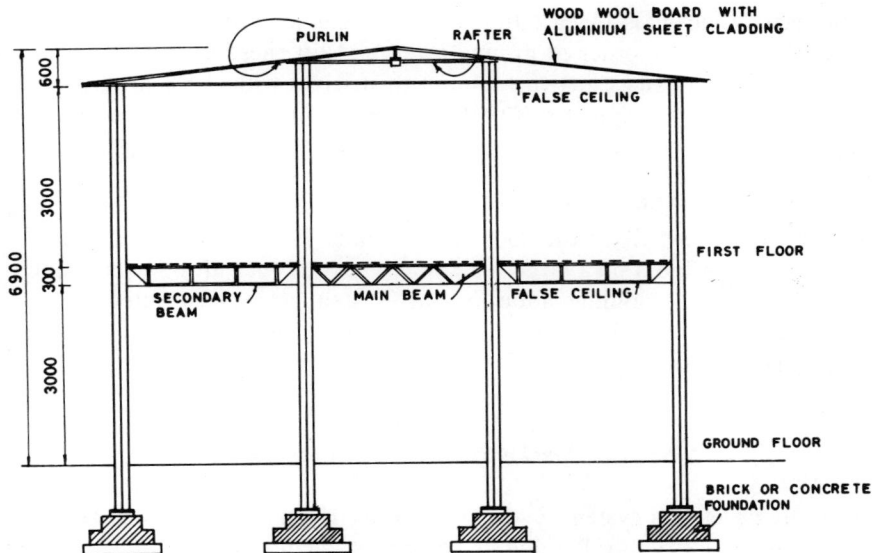

FIGURE 9.5. Cross-section-details of 2-storied structure (super prefab system).

system allows for mass production in a factory of prefabricated components that are easily assembled at the site. The system is specially suited for three categories of houses: deluxe, economy, and low-cost.

The supporting frame consists of prefabricated mild steel columns and beams connected to each other by ms bolts and nuts at the site. The walling is made of chemically treated rice husk and cement concrete. The roof/floor is of conventional cement concrete slab for flat roof/floor and of rice husk or wood wool boards with aluminum cladding for sloped roof. The foundation for the main columns is 12 x 12 ft reinforced cement concrete footing unto which the columns are connected either by bolting or concrete. Figure 9.5 shows a typical section of the building adopting this system.

Lift Slab System

This system is designed to overcome the disadvantages of insite construction that demands a large amount of centering and form work, which in turn delays the construction work. The prefabrication method, an alternate and easy method to insite construction, also has some disadvantages—such as the use of heavy lifting equipment, transport trailers, and so on—that are overcome in the lift slab system.

The basic concept is to cast the horizontal elements in the form of flat plates, one above the other at ground level, and lift these elements to their respective floor heights using the column themselves with the help of jacks. The structural system consists of prefabricated columns in steel or concrete and prestressed concrete slabs to increase stiffness, thereby making it possible to lift them at early ages.

The technique, which is relatively simple and uses unskilled labor, can curtail as much as 50% of the construction time. This system is yet to find its application in India on a large scale for want of proper lifting equipment. However, a model building adopting this system was constructed at the Structural Engineering Research Centre, Roorkee, the feasability of its use for Indian conditions.

3-S System (Shirke Special Structural System)

This newly developed system, by M/S B. G. Shirke and Co., Poona, adopts the concept of a structural frame using precast members and has yet to be used on a large scale. The system is designed to avoid expensive centering for slab and side shuttering for columns.

The supporting structure consists of hollow columns of high-strength concrete designed for all forces, including handling stresses and precast wall panels of suitable type. The flooring/roofing system consists of partially precast beams with flanges designed to support precast slabs. The slab strips may be of either aerated cellular concrete or conventional concrete. The filling-in of the hollow cores of the column and the completion of the top portion of the partially precast beams are done on site. Joints are made monolithic by means of continuity bars provided in the columns and beams.

Large Clay-Block System

Developed by Structural Engineering Research Centre, this system uses precast large wall panels made of hollow clay blocks and reinforced concrete ribs for exterior walls, reinforced concrete panels for interior walls, and waffle shell panels for roofs/floors. Precast waffle shells are used in roof/floor panels. The panels are detailed in such a way as to make insite jointing easy onsite. The vertical joints in between wall panels and horizontal joints between roof/floor panels are made effective by introducing reinforcing steel and concrete to provide overall rigidity to the structure. The system is very much suitable for multistoried residential buildings. Tests have been conducted on wall panels and roof/floor panels at SERC to assess the strength and behavior. Tests on typical horizontal and vertical joints have also been conducted to study the system's behavior under various loadings encountered during and after construction. The series of experiments conducted revealed satisfactory behavior of the system. The advantage of the system could be fully derived by using a part fly ash in concrete to replace a part of expensive cement.

The Tamil Nadu Housing Board is adopting the system for the construction of 144 middle-income group flats at Besant Nagar, Madras.

10

Housing in Venezuela: Case Study

Fernando Tinoco

The problem of housing in Venezuela has been met in the past with solutions that only satisfied the physical appearance of the units without due regard to the cost and capacity of payment of the lower-income groups. Thus the housing needs of these groups were not met and the result was the invasion of lots and the construction of totally inadequate housing units, better known as "ranchos."

The housing shortage and the inadequate living conditions of the ranchos are partly caused by the underdevelopment of the country and the lack of an appropriate national policy. These are the main factors affecting the housing shortage:

1. An incompatibility exists between housing costs and per capita income.
2. The housing construction industry is poorly organized.
3. There are not enough private stimuli to private initiative.
4. There is an absence of appropriate legislation.
5. Proper consideration in planning and budgeting at the nationwide level is lacking.

The incompatibility between housing costs and per capita income plays an important role in the selection, among several possibilities, of a solution to the housing shortages.

The magnitude of the problem is shown in Table 10.1.

TABLE 10.1. Comparison of Family Income to Housing Shortage

Family Income ($/month)	Total Number of Families (%)	Housing Shortage (%)
More than $680.00	10	1
$343.00–$680.00	19	6
$227.00–$342.00	17	11
$114.00–$226.00	31	37
$ 68.00–$113.00	12	22
Less than $ 67.00	11	23

Table 10.1 shows that 23% of the families have an income of less than $115.00 per month, and the housing shortage covers 45% of this group. Furthermore, families with an income of less than $226.00 are 54% of the total number of families, and their housing shortage is 82% of the total shortage. Thus the lower-income groups have the greatest need for appropriate housing. These families in the lower-income groups live in ranchos, which have created social and political problems because of poor living conditions, promiscuity, and a lack of family stability.

Caracas, the capital of Venezuela, presents the problem of the spontaneous settlements that have grown beyond any urbanistic pattern because of the lack of appropriate solutions for dealing with the housing needs of families in the lower-income groups. These settlements, better known as "barrios," are formed by ranchos constructed in a random pattern. The ranchos are being slowly transformed into more adequate housing units, with the increase in the economic capability of the inhabitants of the settlement, by the use of construction materials such as tiles, cement, zinc, steel beams, and so on.

In spite of improvements in the rancho as a housing unit, the settlement still constitutes a very inadequate community because of the lack of water, electricity, sewage and garbage disposal, schools, playgrounds, libraries, and so on. Thus it is necessary to consider solutions that effectively incorporate the barrios with the urbanized living of the city. This can be achieved by taking advantage of the self-improvement capacity of their inhabitants. The sought-after solutions should also accomplish the integration of the inhabitants to the development of the city.

The transformation and organization of a space that had been consolidated chaotically is a challenge. Such a task requires the selection of areas for schools, medical care centers, playgrounds, community centers, roads, and so on, that are now occupied by ranchos. The space need becomes a complex relocation problem for the people living in the barrios, because they may have to move to adjacent areas nearby. However, it must be clear that relocation outside the barrio is not considered possible, with a few exceptions, because of the breakup of the inhabitants' social and neighborhood ties and their resistance to a move that will change their way of living.

Obviously, the transformation of the barrios cannot only be a change in its structure and physical appearance, but there must be a change in the social and cultural attitudes of the inhabitants of the rancho. Thus the participation of the people in the construction of the facilities, the design decision, and all matters related to living in the barrio is a very important aspect of the solution to lower-cost housing.

The construction for lower-cost housing can, but should not, be limited to the appropriate functioning and physical appearance of the housing unit. It must consider the integration of the units within the settlement or barrio, the integration of the barrio with the city, and the integration of the barrio's inhabitants to a better living style. Thus construction of lower-cost housing means building up an environment that may lead to an improved way of living.

A Solution

The national housing problem will definitely be solved with the development of the economical, education, and human resources of the country. However, the families living in ranchos are in urgent need of solutions suitable for immediate application. This is especially true in Caracas, where families with the lowest incomes are forced to live on the steep hills that surround the Caracas Valley. The ranchos built in these hills obstruct natural drainage areas; and with no provision for sanitary facilities, there is a constant leakage of sewer waters that saturate the soil and rock underneath the hills, producing slides during the period of rains with the subsequent losses of effort and lives. In this environment lives a population coming from different parts of the country, with different educational levels, traditions, and social customs and still not adapted to the urban way of life. A quite generalized custom is to use better construction materials whenever the financial situation is improved and also to build a second and third floor to the initial rancho.

The Banco Obrero, a government agency, has constructed different types of solutions to satisfy the housing needs of families in the same income groups. One of the solutions adopted for the lower-income groups is the construction of low-cost single housing units on the steep hills. The construction of these units are limited to slopes with a maximum of 70% of grade, minimizing the amount of soil removal and allowing the use of construction methods of easy execution that could take advantage of the unskilled labor available in the area. Thus the construction is adapted as much as possible to the topography and geological conditions of the site. This solution considers these factos:

1. Full use of the human and financial resources available to these families.
2. The construction of an environment that stimulated the desire for higher educational and economical levels.
3. A construction method of easy execution to be appropriate for the unskilled labor available.
4. A flexible design to allow house improvements by the owner.

Construction Method

The steep slopes are cut in series of successive vertical berms 2 m high by 3.40 m wide. The construction of the berms should avoid excessive removal of soil and, if possible, the stripping of the topsoil. Removing large quantities of soil will increase the cost of construction considerably.

The system used in the construction of the housing unit consisted of load-bearing walls with concrete slabs 2.5 in. thick reinforced with wire mesh and supported by steel beams. This method of construction is well known by the unskilled labor in the area, and the people living in the ranchos have used it to strengthen and enlarge their housing units.

Each unit is built independently of the neighboring units to allow future construction and enlargement of the same. This structural independence allows the flexibility required to adapt it horizontally and vertically to the variable topography encountered and to avoid the removal of large soil quantities.

There has been some experimentation with the mass production of this type of unit. A sliding metal framework is used that allows the construction of a unit per day per set of metal framework.

The Housing Unit

The unit occupies a lot with an area of 48 m^2 and is built in stages to minimize the cost to the family and allow further improvement of the unit by the owner. The Banco Obrero provides the engineering know-how in the construction of the berms, plus the construction of facilities that provide water, electricity and sewage and garbage disposal. This first stage allows the owner to have a starting point for future construction of the unit on his or her own. The result of the second stage is a unit with 38 m² and a capacity for a family of six to nine persons. It provides separate sleeping facilities for parents and children. The children are separated by sex, with the use of bunk beds.

The foundation of the unit is built to permit the addition of a second floor to increase the area up to a total of 76 m². The construction of a second floor constitutes the third and last stage, with the owner deciding whether such construction is needed.

Engineering Considerations

The housing units built on steep slopes required difficult engineering decisions related to the stability and permanency of the berms. They also required decisions about whether to remove the topsoil.

Construction of lower-cost housing does not require low-cost engineering or any engineering at all, because the great part of the cost is absorbed by land costs. Thus construction of lower-cost housing is generally made on difficult foundation conditions that require the best engineering effort. The same could be said about the selection of the structure, materials, and constructive system to be used.

References

Fathy, Hassan. *Architecture for the Poor.* Chicago, Ill. The University of Chicago Press. 1973.

George, Zacharia. *Prefabrication Techniques in India.* Madras, India. Structural Engineering Research Center. December 1974.

Lazar, Benjamin E. *Education and Training of Professionals for the Pre-fabricated Systems Industry.* Proc. of the Third International Symposium on Lower-Cost Housing Problems, Montreal, Canada. Sir George Williams University. 1974.

Nunnally, S. W., "Reducing Florida's Building Costs Through Building Code Reform," Special Report #731, Gainesville, Florida, October, 1973.

Tan, S. H. *Self-Help Housing—Only a Partial Solution.* Proc. of International Conference on Low Income Housing—Technology and Policy. Bangkok, Thailand. Asian Institute of Technology. 1977.

Terzaghi, Karl, and Peck, Ralph B. *Soil Mechanics in Engineering Practice.* (2nd ed.) New York. John Wiley & Sons. 1974. Chapter 2, p. 31.

U.S. Department of Housing and Urban Development. *Aided Self-Help in Housing Improvement.* Ideas and Methods Exchange No. 18. Washington, D.C. Office of International Affairs. 1967.

U.S. Department of Housing and Urban Development. *Aided Self-Help Housing in Africa,* Ideas and Methods Exchange No. 65. Washington, D.C. Office of Internaional Affairs. June 1969.

U.S. Department of Housing and Urban Development. *Aided Self-Help Housing.* Washington, D.C. Office of International Affairs. June 1970.

U.S. Department of Housing and Urban Development. *Industrialized Housing,* Ideas and Methods Exchange No. 66. Washington, D.C. Office of International Affairs. January 1972.

U.S. Department of Housing and Urban Development. *Special Report on Techniques of Aided Self-Help Housing.* Washington, D.C. Office of International Affairs. November 1973.

U.S. Department of Housing and Urban Development. *Earth for Homes.* Ideas and Methods Exchange No. 22. Washington, D.C. Office of International Affairs. October 1974.

Bibliography

Abrams, Charles. *The Future of Housing.* New York. Harper and Brothers. 1946.

Abrams, Charles. *Man's Struggle for Shelter in an Urbanizing World.* (Also *Housing in the Modern World.*) Cambridge, Mass. MIT Press. 1964.

Abrams, Charles. *The Need for Training and Education for Housing and Planning.* New York. United Nations. 1955. (Prepared for the government of Turkey.)

Alonso, William. *Innovations in Housing Design and Construction Techniques as Applied to Low-Cost Housing.* Springfield, Va. National Technical Information Services. 1969.

Anderson, Bruce. *Solar Energy and Shelter Design.* New York. McGraw-Hill Book Co. 1973.

Anderson, L. O. *Low-Cost Wood Homes for Rural America.* Washington, D.C. U.S.A. Forest Service, U.S.D.A. Handbook No. 364. 1969.

Andrzejewski, A. and Kucharski, M. *Financing of Housing in Socialist Countries.* Warsaw, Poland. (Working Paper for the U.N. Third Advisory Group Meeting on Housing Finance.) 1969.

Aronin, J. E. *Climate & Architecture.* New York. Reinhold Publishing Corp. 1953.

Back, Kurt W. *Slums, Projects and People: Social-Psychological Problems of Relocation in Puerto Rico.* Durham, N.C. Duke University Press. 1962.

Baranson, Jack. *Technology for Underdeveloped Areas: An Annotated Bibliography.* Oxford, England. Pergamon Press, 1967.

Barr, Charles W. *Housing-Health Relationships: An Annotated Bibliography.* Monticello, Ill. Council of Planning Librarians, Exchange Bibliography No. 82. 1969.

Bauer, Catherine. *Social Questions in Housing and Town Planning.* London, England. University of London Press, Ltd. 1952.

Becker, Franklin D. *Housing Messages.* University of Virginia. 1977. New York. McGraw-Hill Book Co.

Bell, Gwendolyn D. *Urban Environments and Human Behavior.* New York. McGraw-Hill Book Co. 1973.

Bemis, Albert F. *Rational Design, Volume III, The Evolving House.* Cambridge, Mass. MIT Press. 1933.

Bemis, Albert F. *The Evolving House.* Cambridge, Mass. The Technology Press, MIT. 1936.

Benson, Ben. *Critical Path Methods in Building Construction.* Englewood Cliffs, N.J. Prentice-Hall. 1970.

Blachere, G. *Savoir Batir, Habitabilite, Durabilite, Economie* Des Batiments. Paris, France. Editions Eyrolles. 1966.

Boice, John R. *A History and Evaluation of SCSD, Building Systems.* Menlo Park, Calif., California Information Clearing House.

Branch, Meville C. *Planning Urban Environment.* Los Angeles, Calif. University of Southern California. 1975. New York. McGraw-Hill Book Co.

Breese, Gerald. *The City in Newly Developing Countries.* Englewood, N.J. Prentice Hall. 1969.

Brinkworth, Brian J. *Solar Energy for Man.* New York. Halsted Press Book. John Wiley and Sons, Inc. 1972.

Brunn, Stanley D. *Urbanization in Developing Countries: An International Bibliography.* (Report #8.) East Lansing, Mich. Michigan State University. 1971.

Building Research Advisory Board. *Foundations for Residential Structures in Seismic Areas.* Washington, D.C. National Academy of Sciences. 1969.

Burberry, Peter. *Building for Energy Conservation.* Somerset, N.J. John Wiley & Sons, Inc. 1978.

Burg, Nan C. *Rural Housing and Rural Poverty: A Bibliography.* Monticello, Ill. Council of Planning Librarians, Exchange Bibliography #247. 1971.

Burnham, Kelly. *The Prefabrication of Houses.* Cambridge, Mass. MIT Press. New York. John Wiley & Sons, Inc. 1951.

Burns, Leland S. *Housing: Symbol and Shelter.* Los Angeles, Calif. University of California. 1970.

Bush, Vincent G. *Construction Management: A Handbook for Contractors, Architects & Students.* Englewood Cliffs, N.J. Prentice-Hall. 1973.

Calderwood, D. M. *Principles of Mass Housing.* Johannesburg, South Africa. Cape and Transvaal Printers. 1964.

Callaway, Thomas R. *Housing in the Ivory Coast.* Washington, D.C. HUD., Division of International Affairs, Country Report Series. 1966.

Caminos, H., Turner, J. F. C., and Steffian, John. *Urban Dwelling Environments—An Elementary Survey of Settlements for the Study of Design Determinants.* Cambridge, Mass. MIT Report 16. MIT Press. 1969.

Clinars, Marshall B. *Slums and Community Development: Experiments in Self-Help.* New York. Free Press. 1966.

Conway, Donald J. *Human Response to Tall Buildings.* American Institute of Architects. 1974. New York. McGraw-Hill Book Co. 1977.

Corner, Eric. *Modular Primer.* London, England. Modular Society Ltd. 1963.

Cowan, Henry J. *Science and Building: Structural and Environmental Design in the Nineteenth and Twentieth Centuries.* Somerset, N.J. John Wiley & Sons, Inc. 1978.

Crane, Theodore. *Architectural Construction: The Choice of Structural Design.* (2nd ed.) Somerset, N.J. John Wiley & Sons, Inc. 1956.

Cutler, Laurence S. *Industrialized Building Systems.* (Tape & Cassett.) Hightstown, N.J. McGraw-Hill Book Co. 1970.

Dakhil, Fahd H., Ural, Oktay, and Tewfik, Moneer F. *Housing Problems in Developing Countries*. Vol. 1. New York. John Wiley & Sons, Inc. 1978.

Deeson, A. F. L. *The Comprehensive Industrialized Building Systems Annual—1967*. Kent, England. Product Journals Ltd. 1967.

Deeson, A. F. L. *The Comprehensive Industrialized Building Systems Annual—1970*. London, England. Morgan-Grampion. 1970.

de Neufville, Richard, and Marks, David H. *Systems Planning and Design: Case Studies in Modelling, Optimization, and Evaluation*. Englewood Cliffs, N.J. Prentice-Hall. 1974.

Diamant, R. M. E. *Industrialized Building* (Three volumes.) London, England. Lliffe Books, Ltd. 1964.

Diamant, R. M. E. *Industrialized Buildings, Volumes I, II, III. The Architect and Building News*. London, England. Lliffe Books, Ltd. 1968.

Dietz, A. G. H., and Cutler, Laurence S. *Industrialized Building Systems for Housing*. Cambridge, Mass. MIT Press. 1971.

Dodge, J. Robert, Young, Burton O., and Wilson, L. Albert. *Housing in Ethiopia*. Washington, D.C. HUD. Office of International Housing, Country Report Series. 1966.

Doxiadis Associates. *Housing in Libya*. Athens, Greece. Prepared for the Ministry of Planning and Development by Doxiadis Associates. 1964.

Doxiadis, Konstantinos A. *Architecture in Transition*. London, England. Oxford University Press. 1963.

Doxiadis, Konstantinos A. *Ekistics: An Introduction to the Science of Human Settlements*. New York. Oxford University Press. 1968.

Drew, Jane B., and Fry, E. Maxwell. *Tropical Architecture in the Dry and Humid Zones*. New York. Reinhold Publishing Corp. 1964.

Eastman, Charles M. *Spatial Synthesis in Computer-Aided Building Design*. Somerset, N.J. John Wiley & Sons, Inc. 1975.

Egan, M. D. *Concepts in Building Firesafety*. Somerset, N.J. John Wiley & Sons, Inc. 1978.

Elliott, Sean H. *Financing Latin American Housing; Domestic Savings Mobilization and U.S. Assistance Policy*. New York. Praeger. 1968.

Engelmann, Konrad. *Building Cooperative Movements in Developing Countries: The Sociological and Psychological Aspects*. New York. Praeger. 1968.

Ettinger, Jan van. *Towards a Habitable World; Task-Problems and Methods-Acceleration*. New York. (Published for Bouwcentrum, Rotterdam.) Elsevier Publishing Co. Amsterdam. Holland. 1960.

Ettinger, Jan van. *Problems and Methods of Low-Cost Housing*. Bouwcentrum, Rotterdam. Holland. 1969.

Europrefab Systems Handbook, Housing. London, England. Interbuild Prefabrication Publications, Ltd. 1969.

Evans, Hazel. *New Towns: The British Experience*. Somerset, N.J. John Wiley & Sons, Inc. 1972.

Fareed, Adel, El-Hifnawi, Mostafa, and Ural, Oktay. *IAHS Cairo Workshop on Evaluation of Industrialized Housing Systems*. November 15-20, 1976. Cairo, Egypt. IAHS Cairo Regional Office.

Fazio, Paul et al. *Third International Symposium on Lower Cost Housing Problems*. May 27-31, 1974. Rolla, Mo. Concord University, Systems Building Center.

Bibliography

Galbraith, John K. *The New Industrial State.* Boston, Mass. Houghton Mifflin Company. 1967.

Golany, Gideon. *New-Town Planning: Principles and Practice.* Somerset, N.J. John Wiley & Sons, Inc. 1976.

Golany, Gideon. *International Urban Growth Policies: New Town Contributions.* Somerset, N.J. John Wiley & Sons, Inc. 1978.

Habraken, N. J. *Supports: An Alternative To Mass Housing.* London. The Architectural Press. 1972a.

Habraken, N. J. *Supports: An Alternative to Mass Housing.* New York. Praeger. 1972b.

Hamilton, Richard. *Space Heating with Solar Energy.* MIT. August 20–26, 1950. New York. McGraw-Hill Book Co.

Harrell, Raymon H., and Lendrum, James. *A Demonstration of New Techniques for Low-Cost Small Home Construction.* Washington, D.C. HUD (Housing and Home Finance Agency. Housing Research Paper No. 29. 1954.

Harrison, H. W. *Performance Specifications for Building Components.* Watford, England. Ministry of Public Building and Works. 1969.

Hart, Franz, Henn, Walter, and Sonntag, Hansjurgen. *Multi-Storey Buildings in Steel.* Somerset, N.J. John Wiley & Sons, Inc. 1978.

Hesselgren, Sven. *Man's Perception of Man-Made Environment.* Stockholm, Sweden. Royal Institute of Technology. 1976. New York. McGraw-Hill Book Co.

Hester, Randolph T., Jr. *Neighborhood Space: User Needs and Design Responsibility.* Raleigh, N.C. North Carolina State University. 1976. New York. McGraw-Hill Book Co. 1976.

Hoffman, Hubert. *Row Houses and Cluster Houses: An International Survey.* New York. Praeger. 1957.

Holliday, John. *City Centre Redevelopment.* Somerset, N.J. John Wiley & Sons, Inc. 1973.

Hornbostel, Caleb. *Construction Materials: Types, Uses and Applications.* Somerset, N.J. John Wiley & Sons, Inc. 1978.

Hornbostel, Caleb, and Hornung, William J. *Materials and Methods for Contemporary Construction.* Englewood Cliffs, N.J. Prentice-Hall. 1974.

Hosken, Fran P. *The Functions of Cities.* Cambridge, Mass. Schenkman Publishing Co. 1973.

Housing Construction. Moscow, Russia. Novosti Press Agency Publishing House. 1967.

International Council for Building Research, Studies, and Documentation. *Towards Industrialized Building.* Copenhagen, Amsterdam. 3rd CIB Congress. Elsevier Publishing Co. 1966.

International Labor Office. *Workers' Housing Problems in Asian Countries.* Asian Regional Conference, Tokyo, Japan. Report II, International Labor Organization, Geneva, Switzerland. 1953.

International Labor Office. *Housing Cooperatives.* Geneva, Switzerland. ITS Studies and Reports No. 66. 1964.

Kelly, Burnham. *The Prefabrication of Houses.* Cambridge Mass. The Technology Press. MIT. 1951.

Kelly, Phyllis M. *International Bibliography of Prefabricated Housing.* Cambridge, Mass. MIT. 1954.

Knudson, Vern O., and Harris, Cyril M. *Acoustical Designing in Architecture.* Somerset, N.J. John Wiley & Sons, Inc. 1950.

Koncelik, Joseph A. *Designing the Open Nursing Home.* Columbus, Ohio. Ohio State University. 1976. New York. McGraw-Hill Book Co.

Lang, Jon, Burnette, Charles, Moleski, Walter, and Vachon, David. *Designing for Human Behavior.* New York. McGraw-Hill Book Co. 1975.

La Patra, Jack W. *Applying the Systems Approach to Urban Development.* University of California. Davis, California. 1973. 1977. New York. McGraw-Hill Book Co.

Lewicki, Bohdan. *Building with Large Prefabricates.* New York. Elsevier Publication Co. 1966.

Lewis, David. *The Growth of Cities.* Somerset, N.J. John Wiley & Sons, Inc. 1971.

McLuhan, Herbert M. Understanding Media. *The Extensions of Man.* New York. McGraw-Hill Book Co. 1964.

Macsai, John, Holland, Eugene P., Nachman, Harry S., and Yacker, Julius Y. *Housing.* Somerset, N.J. John Wiley & Sons, Inc. 1976.

Mainder, W. F. *Hong Kong Urban Rent and Housing.* Hong Kong. Hong Kong University Press. 1969.

Malmstrom, P. E., and Munch-Peterson, J. C. *Philosophy of Design of Industrialized Housing.* U.N. Social and Economic Council ST/ELLA/CONF. (27/L,5.) 1967.

kus, Thomas A., *Building Performance.* Somerset, N.J. John Wiley & Sons, Inc. 1972.

Markus, Thomas A., and Morris, Edwin N. *Buildings, Climate and Energy.* Somerset, N.J. John Wiley & Sons, Inc. 1978.

Martin, Bruce. *The Coordination of Dimensions for Building.* London, England. RIBA. 1965.

Meyerson, M., Terret, B., Wheaton, W. L. C. *Housing, People and Cities.* New York. McGraw-Hill Book Co. 1962.

Miller, J. Marshall. *New Life for Cities Around the World—International Handbook on Urban Renewal.* New York. Books International. 1959.

Misra, Surya K. *Building and Planning in Developing Nations.* Stockholm, Sweden. National Swedish Institute for Building Research. 1967.

Modular Coordination in Building. New York. United Nations. 1968.

Modular Coordination of Low-Cost Housing. New York. United Nations. 1966.

Moore, Gary T., and Golledge, Reginald G. *Environmental Knowing.* New York. McGraw-Hill Book Co. 1976.

Muth, Richard F. *Cities and Housing.* Chicago, Ill. University of Chicago Press. 1969.

Newmark, Nathan M., and Rosenblueth, Emilio. *Fundamentals of Earthquake Engineering.* Englewood Cliffs, N.J. Prentice-Hall. 1971.

Nijhoff, Martinus. *Urbanization in Developing Countries.* The Hague. International Union of Local Authorities. 1968.

Oakley, David. *The Phenomenon of Architecture in Cultures in Change.* New York. Pergamon Press. 1970.

Oldman, Oliver. *Financing Urban Development in Mexico City: A Case Study of Property Tax, Land Use, Housing and Urban Planning.* Cambridge, Mass. Harvard University Press. 1967.

Olgyay, Victor. *Design with Climate: Bioclimatic Approach to Architectural Regionalism.* Princeton, N.J. Princeton University Press. 1963.

Oliver, Paul. *Shelter and Society.* New York. Praeger. 1969.

Organization for Economic Cooperation and Development. *The Role of Trade Unions in*

Housing. Paris, France. International Seminars 1967, Final Report, Manpower and Social Affairs Directorate. 1968.

Pama, R. P., Angel, S., and De Goede, J. H. *International Conference on Low-Income Housing—Technology and Policy.* June 7–10, 1977. Bangkok, Thailand. Asian Institute of Technology.

Parameswaran, V. S. *International Seminar on Low Cost Housing.* January 12–22, 1977. Adyar, Madras, India. Organizing Secretary SERC, Council for Scientific and Industrial Research Campus.

Parvis, F. Rad, Busching, Herbert W., and Ural, Oktay. *Fourth IAHS International Symposium on Housing Problems.* May 24, 28, 1976. Elmsford, New York. Pergamon Press.

Pawley, Martin. *Architecture vs. Housing.* New York. Praeger. 1971.

Peattie, Lisa. *Slums.* Monticello, Ill. Council of Planning Librarians, Exchange Bibliography No. 113. 1970.

Plesums, Guntis. *Townframe: Environments for Adaptive Housing.* Eugene, Oregon. University of Oregon. 1978. Stroudsburg, Pa. Dowden, Hutchinson & Ross, Inc.

Pollowy, Anne-Marie. *The Urban Nest.* Canada. University of Montreal. 1977. New York. McGraw-Hill Book Co.

Popko, Edward. *Transitions: A Photographic Documentary of Squatter Settlements.* New York. McGraw-Hill Book Co. 1978.

Rapport, Amos. *House Form and Culture.* Englewood Cliffs, N.J. Prentice-Hall. 1969.

The Report of the President's Committee on Urban Housing (The Kaiser Committee). *A Decent Home.* Washington, D.C. 1968.

Rodwin, Lloyd. *Urban Planning in Developing Countries.* Washington, D.C. Department of Housing and Urban Development. Ideas and Methods Exchange No. 61. 1965.

Rosen, Sherman J. *Manual for Environmental Impact Evaluation.* Englewood Cliffs, N.J. Prentice-Hall. 1976.

Safdie, Moshe. *Beyond Habitat.* Cambridge, Mass. MIT Press. 1970.

Sanoff, Henry. *A Bibliography and Critical Review of Industrialized Housing.* Monticello, Ill. Council of Planning Librarians, Exchange Bibliography No. 158. 1970.

Sanoff, Henry. *Methods of Architectural Programming.* Raleigh, N.C. North Carolina State University. 1977. New York. McGraw-Hill Book Co.

Sanoff, Henry. *Designing with Community Participation.* Raleigh, N.C. North Carolina State University. 1978. New York. McGraw-Hill Book Co. 1978.

Schmid, Thomas, and Testa, Carlo. *Systems Building: An International Survey of Methods.* London, England. Pall Mall Press. 1969.

Seaton, Richard W. *Social Factors in Architectural and Urban Design.* Monticello, Ill. Council of Planning Librarians, Exchange Bibliography No. 201. 1971.

Sebestyen, Gyula. *Large Panel Buildings.* Budapest, Hungary. Akademiai Kiado, 1965.

Sebestyen, Gyula. *Large Panel Buildings.* Budapest, Hungary. Akademiai Kiado. Hungarian Academy of Sciences. 1967.

Singapore Housing & Development Board. *First Decade in Public Housing.* 1960–1969. Singapore. Times Printers. 1970.

Slate, Floyd O. *Low-Cost Housing for Developing Countries.* Ithaca, New York. Cornell University. 1974.

Smith, Wallace F. *Housing: The Social and Economic Elements:* Berkeley, Calif. University of California Press. 1970.

Smithson, A. and P. *Housing Primer: Low and Medium Rise Housing.* London, England. Architectural Design, 1967.

Stearns, Forest W., and Montag, Tom. *The Urban Ecosystem: A Holistic Approach.* Madison, Wis. University of Wisconsin, 1975. New York. McGraw-Hill Book Co. 1975.

Testa, Carlo, and Schmid, Thomas. *Systems Building: An International Survey of Methods.* New York. Praeger. 1969.

Thomas, Mark H. *Modular Design of Low-Cost Housing.* New York. United Nations. 1966.

Turner, John F. C., and Fichter, Robert. *Freedom to Build: Dweller Control of the Housing Process.* New York. MacMillan Co. 1972.

U.N. *Cumulative List of United Nations Documents and Publications in the Field of Housing, Building and Planning.* List No. 12. New York. U.N. Publications. 1970.

U.N. *Survey of Current Housing Problems and Government Policies.* New York. 1971.

United Nations. *Proceedings of the United Nations Conference on New Sources of Energy,* Vols. 4–6. Rome, Italy. 1961. New York. McGraw-Hill Book Co.

U.N. Center for Housing, Building and Planning. *Cumulative List of U.N. Documents and Publications Related to the Field of Housing, Building and Planning.* New York. U.N. Publications. 1968.

U.N. Department of Economic and Social Affairs. *World Housing Conditions and Estimated Housing Requirements.* New York. 1965.

U.N. Department of Economic and Social Affairs. Statistical Office. *Statistical Indicators of Housing Conditions.* New York. 1962.

U.N. Department of Economic and Social Affairs. *Rural Housing: A Review of World Conditions.* New York. 1969.

U.N. Economic Commission for Europe. *Housing in the Less Industrialized Countries of Europe.* Geneva, Switzerland. 1956.

U.N. Social Commission. *Long Range Programme of Concerted International Act in the Field of Low-Cost Housing and Related Community Facilities.* New York. 1959.

U.N. Statistical Office. *Principles and Recommendations for the 1970 Housing Census.* New York. 1967.

Ural, Oktay. *First International Symposium on Low Cost Housing Problems, Related to Urban Renewal and Development.* October 8–9, 1970. Rolla, Mo. University of Missouri, Extension Division.

Ural, Oktay. *Second International Symposium on Low Cost Housing Problems.* April 24–26, 1972. Springfield, Va. NTIS National Technical Information Service.

Ural, Oktay. *A Systematic Approach to Basic Utilities in Developing Countries.* May 1974. Washinton, D.C. HUD. Office of International Affairs.

Ural, Oktay, "Political, Economic and Technological Conditions in Different Parts of the World in Relation to the Applicability of Systems Approach to Housing Production," Urban Technology Conference, National Leage of Cities, New York, May 1971.

Ural, Oktay, "Systems Approach and Industrialization in Housing in the Western World," XI IFAWPCA Convention on Low Cost Housing, New Delhi, India, November 1971.

Ural, Oktay. *New Trends on Lower Cost Housing Production, Emphasizing the Problems of Emerging Countries.* May 11–15, 1975. Miami, Florida. Florida International University (International Association for Housing Science, Tamiami Campus).

Ural, Oktay, and Celik, Aliye. *International Conference on Disaster Area Housing.* September 4–10, 1978. Cankaya, Ankara, Turkey. Turkish Building Research Institute.

Ural, Oktay, and Tokman, Bulent. *RCD Lower Cost Housing Workshop.* November 1-4, 1977. Cankaya, Ankara, Turkey. Turkish Building Research Institute.

U.S. Congress, Senate, Committee on Banking and Currency, Subcommittee on Housing, 87th Congress, 2nd Session. *Report on International Housing Programs.* Washington, D.C. U.S. Government Printing Office. 1962.

U.S. Department of Commerce, National Bureau of Standards. *Industrialized Building in the Soviet Union.* Washington, D.C. NBS Special Publications No. 334. May 1972.

U.S. Department of Housing and Urban Development. *Housing in Liberia.* Washington, D.C. Office of International Housing, Country Report Series. No date.

U.S. Department of Housing and Urban Development. *Manual on Design for Low-Cost and Aided Self-Help Housing.* Washington, D.C. Office of International Housing, Ideas and Methods Exchange No. 37. 1957 (reprinted 1961).

U.S. Department of Housing and Urban Development. *Prefabricated Concrete Components for Low-Cost Housing Construction.* Washington, D.C. Office of International Affairs, Ideas and Methods Exchange No. 59. 1963.

U.S. Department of Housing and Urban Development. *Housing in Guatemala.* Washington, D.C. Office of International Housing, Country Report Series. 1965a.

U.S. Department of Housing and Urban Development. *Housing in Nigeria.* Washington, D.C. Office of International Housing, Country Report Series. 1965b.

U.S. Department of Housing and Urban Development. *Aided Self-Help in Housing Improvement.* Washington, D.C. (Prepared for USAID.) Office of International Affairs, Ideas and Methods Exchange No. 18. 1967.

U.S. Department of Housing and Urban Development. *A Comparative Analysis of European Experience.* Washington, D.C. Office of International Affairs, Special Report, Industrialized Building. 1968a.

U.S. Department of Housing and Urban Development. *Manual for Low-Cost and Aided Self-Help.* Washington, D.C. Office of International Affairs. 1968b.

U.S. Department of Housing and Urban Development. *Operation Breakthrough: Housing, a Bibliography.* Ithaca, New York. Cornell University. 1970.

U.S. Department of Housing and Urban Development. *Housing and Urban Development in Japan.* Washington, D.C. HUD. Office of International Affairs, International Brief. 1971a.

U.S. Department of Housing and Urban Development. *Housing Systems Proposals for Operation Breakthrough.* Washington, D.C. Government Printing Office. 1971b.

U.S. Department of Housing and Urban Development. *Panama.* International Country Report, Office of International Affairs. 1971c.

U.S. Department of Housing and Urban Development. *Peru.* International Country Report, Office of International Affairs. 1971d.

U.S. Department of Housing and Urban Development and MITRE. *An Analysis of Twelve Experimental Housing Projects.* U.S. Government Printing Office, ST/MI. 1966.

Van Huyck, Alfred P. *Planning for Sites and Services Programs.* Washington, D.C. HUD. Office of International Affairs, Ideas and Methods Exchange No. 68, 1971.

Varsheya, J. K. and Mathur, G. C. *Handbook of Rural Housing and Village Planning.* New Delhi, India. National Building Organization. 1968.

Wagner, Bernard. *Housing in Jordan.* Washington, D.C. HUD. Office of International *Housing.* 1965.

Walker, Charles R. *Technology, Industry and Man in the Age of Acceleration.* New York. McGraw-Hill Book Co. 1968.

West, Herbert W. H. *The Establishment of the Brick and Tile Industry in Developing Countries.* New York. United Nations. 1969.

Winne, Robert. *Bibliography on Squatter Settlements, Urbanization in Developing Countries.* Troy, New York. Rensselaer Polytechnic Institute. 1969.

Wolfskill, Lyle A., Dunlop, Wayne A., and Callaway, Bob M. *Handbook for Building Homes of Earth.* Washington, D.C. HUD. Office of International Affairs. No date.

Zarem, A. M., and Erway, Duane D. *Introduction to the Utilization of Solar Energy.* New York. McGraw-Hill Book Co. 1963.

Index

Advantages and disadvantages of industrialized construction, 151
  construction technology, 153
  heavy concrete module, 153
  housing market, 151
  industrialized construction, 152
  industrialized countries, 151
  manufacturing process, 153
  mobile home industry, 151
  performance code, 152
  wood panelized homes, 152
Aided self-help housing, 230
  aided self-help building, 232, 234
  infrastructural facilities, 235
  implementation, 235
  population, 230

Bearing wall construction, 51
  bonding, 56
  buckling, 55
  crushing, 55
  gravity bearing walls, 51
  lateral support, 57
  load-bearing, 57
  load distribution, 55
  pole, structures, 58
  steel reinforced bearing walls, 51
  thickness, 57
  uniform loading, 57
  wood frame construction, 57
Building materials, 33
  concrete, 34
  cracking, 34
  dimensional stability, 34
  industrialized housing construction, 34
  insulating properties, 34
  metals, 34
  plastic, 34
  wood, 34

Building systems software, 154
  world population, 154

Classical construction, 35
  "chaussee", 35
  low-costing housing, 36
  Marcus Vitruvius, 35
  Romans, 35
Cooking, 115
  energy crisis, 115
  heating, 116
  hibachi, 116
  solar ovens, 115, 116

Decision making, 166
Degrees of industrialization, 124
  Agency for International Development, 129
  building codes, 125
  capital improvements, 129
  components, 124
  Department of Labor, 129
  double-wide mobile homes, 126
  Federal Home Loan Bank Board, 126
  flat roofs, 132
  glass fiber, 130
  glazed tile, 131
  gypsum, 125
  homebuilders, 124
  industrialization, 124, 135
  industrialized building, 137
  industrialized housing, 130
  industrialized systems, 126
  interest rates, 126
  land speculation, 126
  Latin America, 137
  manufacturing plant, 133
  median cost, 125
  mobile homes, 125, 126, 130, 132
    cost, 134

272 Index

number, 135
modular, 131
modular units, 132
mortgage, 127
multifamily housing, 126
National Association of Home Builders, 124
onsite labor, 126, 130
Operation Breakthrough, 126
population growth, 137
prefabricated building, 137
prefabricated homes, 125, 131
prefabrication, 132
prehung doors, 131
ready-mix concrete, 125
rehabilitation, 135
scheduling, 131
semi-industrialization, 126, 128, 129, 130, 135, 138
standardized components, 135
transportation, 128
Ucopan system, 125
unskilled labor, 129
zoning, 125
zoning ordinances, 138

Energy, 36
photosynthesis, 36

Financial control, 161
systems builder, 161
Fire, 81
combustion, principles of, 83
fire prevention, 84
Great Fire of London, 82
King Louis XIV, 82
London Rebuilding Act, 82
masonary walls, 82
plank frame construction, 85
plaster, 84
plasterboard, 84
plaster of paris, 82
Flat roofs, 95
asphaltic membranes, 97
drainage of, 96
Foundations and footings, 75
clay, 76
cohesive mass, 76
end-bearing piles, 81
piles, 80
plastic soils, 79
Portland cement, 78

rafts, 79
soils, 75
spread footings, 78
stone, 76
Tower of Pisa, 76
Venice, 81

Government action, 10
closed systems, 12
decision process, 14
industrialization, 14
infrastructures, 14
legislation, 14
long-term loans, 12
minimum-space allocation, 12
national building research institutes, 11
national development program, 11
national housing policy, 14
open systems, 11
Portland cement, 15

Heat and humidity, 106
adobe, 106
air conditioning, 108
dry climates, 106
evaporative cooling, 106
hot-humid climates, 108
thin walls, 108
ventilation, 106
Heavy box systems, 222
box system, 222
capital investment, 223
performance evaluation, 229
process of fabrication, 224
tunnel systems, 222, 229
History of industrialized construction, 141
industrialized construction, 144
labor rates, 146
mechanical modules, 146
modules, 140
Operation Breakthrough, 142
open system, 140
panel system, 140
post tension, 140
prefabrication, 140
prestress, 140
productivity, 146
residential occupancy, 144
sectional modular, 141
steel-framed building, 145
subsystems, 141
system, 141

systems building, 141
  wood frame, 145
Horizontal loads, 58
  dynamic loads, 58
  earthquake loads, 58
  earthquakes, 59
  horizontal shear, 60
  tectonic plate, 60
  thatched roof, 59
  wind loads, 58
Housing, 1
  low-cost, 1
  lower-cost of, 1
  no-income groups, 3
Housing systems in India, 245
  Bombay, 249
  castone system, 246
  Ford Foundation, 246
  integral house, 246
  large clay-block system, 252
  lift slab system, 251
  prefabrication method, 251
  Structural Engineering Research Centre, 245
  super prefab system, 249
  Tamil Nadu Housing Board, 253
  "tapsia type", 246
  Ucopan system, 245
  United Nations Development Program, 245

Implementation, 6
  developing countries, 8
  interdisciplinary technical team, 9
  site planning, 8
  Urban Development Program, 6
Indigenous materials, 28
  labor-intensive, 28
  light-gage steel, 23
  masonry, 24
  mortar, 26
  plastics, 26
  pozzolanic materials, 28
  preformed, 27
  properties, 22, 23
  reinforcing steel, 23
  sheet products, 24
  steel, 22
  stone, 24
  thermal conductivity, 22
Industrialized housing, 139
  box system, 139
  closed system, 139
  components, 139
  frame system, 140
  high-density living, 142
Infrastructure, 162
  advance purchase systems, 166
  building codes, 163
  housing cooperatives, 165
  industrialized process, 162
  land, 166
  loan systems, 165
  mortgage and financing, 165
  nationalization of land, 166
  software aspects of building, 165
  software elements, 162
  speculation on land, 166
  standards, 163
  systems building, 162, 165
  zoning and subdivision, 164

Land, 237
  aided self-help housing, 237
  construction method, 239
  financing and foreign aids, 238
  "The German Experiment", 243
  "The Greek Experiment", 242
  group and individual self-help, 237
  labor-intensive, 239
  low-cost rural housing, 240
  "The Moroccan Experiment", 243
  participation of families in construction, 238
  Puerto Rican experiment, 242
  Puerto Rican self-help housing, 238
  rammed earth, 239
  self-help housing, 244
  "The Swedish Experiment", 243
Lateral stability, 60
  cantilevers, 60
  earthquake forces, 66
  masonry walls, 66
  shear walls, 60
  suspension structures, 70
Light and medium weight panels, 180
  Balency, 194
  Ball brothers, 180
  boxes, 201
  concrete wall panels, 194
  fabrication, 195
  government intervention, 193
  heavy panel systems, 194
  Jesperson, 194

Larson & Nielson, 194
materials systems, 181
national homes, 185
Operation Breakthrough, 180, 194
panelfab, 180
performance evaluation, 183, 197
Prepsa, 180
Skarne, 194
tilt-up panels, 195, 197
Tracoba, 194
Union of Soviet Socialist Republics, 192
wood frame panel systems, 181

Management, 160
Marketing, 161
Medium weight box systems, 201
  checkerboard stacking, 221
  chem-stress, 215
  dry-joint paneling, 203
  habitat, 214
  heavy box systems, 211
  industrialized housing, 215
  mechanical systems, 203
  medium weight box system-mobile, 201
  mobile homes, 204
  Moshe Safdie's Habitat, 213
  Operation Breakthrough, 208
  performance evaluation, 211, 214, 218
  production sequence, 208
  sectional homes, 208
  software, 204
  transhelters, 203
  tunnel forms, 218
  wood frame box systems, 201
  Zachry, 215
Metals, 22
  aluminum alloys, 23
  bamboo, 28
  building bricks, 25
  cold-formed, 23
  corrosion, 22
  high polymers, 26
Modular coordination, 166

Noncombustible construction, 112
  low-rise buildings, 114
  post and beam construction, 114

Owner, 160

Panel systems, 176
  heavy weight panel, 179
  lightweight panel, 177
  medium weight panel, 179
People, 9
  high-density development, 9
  mobile homes, 10
  squatter areas, 9
Pitched roofs, 39
  A-frame construction, 47
  corrugated asbestos, 45
  frame systems, 49
  joists, 51
  no-thrust systems, 39
  planks, 51
  roof trusses, 47
  scaffolding, 45
  shear connectors, 49
  systems with horizontal thrust, 39
  "trussed rafters", 49
Planning, 2
  critical path method, 5
  Descartes's, 5
  ICES-Project II, 5
  "success contribution," 5
  systems approach, 3, 5
  "weighing functions," 3
Production control and coordination, 161
Promise of industrialization, 154
  industrialization, 154
  systems building, 155
Properties of concrete, 15, 16
  dimensional stability, 17
  durability, 17
  fire resistance, 18
  insulating qualities, 18
  strength, 17
  thermal conductivity, 18
  thermal expansion, 18
  unit-weight, 18
  workability, 16

Question of cost, 155
  conventional methods, 156
  "cost", 155
  cost factors, 156
  high-rise construction, 155

Research and development, 166
Roof construction, 37
  flat low slope roofs, 37
Roof terraces, 115

Sanitation, 119
  climate, 121
  drinking water, 119
  fresh water, 119
  industrial building systems, 122
  low-cost shelter, 122
  natural light, 121
  plastics, 121
  Romans, 121
Semi-industrialized, 123
  industrialization, 123
  mobile home, 123
  population, 124
  United Nations, 123
Software aspects, 159
  systems building, 159
Soil, 28
  air voids, 31
  classification of, 28
  cohesive soil, 32
  compacted soil, 32
  compaction, 32
  cone test, 32
  definition of, 28
  effective stress, 30
  engineering classification, 29
  foundations of, 33
  granular soils, 31
  Karl Terzaghi, 30
  pedologic soil, 29
  permeability, 30
  plate load tests, 31
  pore pressures, 30
  proctor test, 32
  properties of, 29
  shear resistance, 31
  shear strength of, 31
  shrinking and swelling, 32
  stress strain and compressibililty, 30
Sponsor, 158
  architectural designers, 159
  economists, 159
  legal counselors, 159
  planners, 158
  real estate experts, 159
  sociologists, 158
  system designers, 159
Systems building hardware, 168
  architectural, 171
  box systems, 173
  classification of systems, 169
  "componoform", 174
  concrete frames, 175
  degrees of industrialization, 171
  durastress, 175
  economic and social, 171
  erection, 171
  frame systems, 172, 173
  generic systems hardware, 172
  heavy frame systems, 174
  industrialized construction, 168
  light frame systems, 174
  medium weight frame systems, 174
  "Mitchel system", 174
  modular system, 168
  Operation Breakthrough, 175
  panel system, 168, 173
  parameters, 171
  post and beam system, 168
  Projekton, 175
  Royal Institute of British Architects, 168
  school construction system development, 174
  spancrete, 175
  structural, 171
  transportaton, 171
System design, 157
  systematic approaches, 155
  traditional building, 157

Thatch roofs, 86
  bamboo pegs, 91
  concrete tiles, 92
  corrugated iron, 95
  curved tiles, 93
  Eastern cedar, 91
  Italian, 92
  pitched, 95
  ribbed surface, 93
  shingles, 89
  slates, 89
  sloped roofs, 95
  Spanish, 92
  Tanzania, 88
  thatch, 88
  tiles, 92
Training programs, 235
Types of industrialized construction, 147
  balency system, 150
  box system, 147
  echo-modular, 150
  "horizontally postured", 147
  Larsen-Nielson, 150
  mobile home, 147

276 Index

school construction system development, 147
Thamesmead Project, 150
"vertically postured", 147

Urbanized land, 5
 Columbia City, 6
 infrastructures, 5
 land speculations, 5
 low-cost housing, 6

Vaults and domes, 70
 catenary arches, 72
 parabolic arches, 72, 75
 sun-dried bricks, 74
 ventilation, 74
Venezuela case study, 254
 Banco Obrero, 256, 258
 barrio, 256
 Caracas, 255
 construction method, 257
 housing unit, 257
 lower-cost housing, 256
 "ranchos", 254
 steep slopes, 258
 urbanized living, 255

Walls, 97
 coatings, 98
 insulation, 101
 joints, 101
 latex paints, 98
 masonry walls, 97
 porous materials, 102
 shrinkage of, 98
 terra cotta blocks, 99
Wood, 102, 110
 brick, 103
 "chinking", 102
 diffusion, 105
 frame construction, 110
 lime mortar, 102
 log construction, 110
 masonry construction, 103
 masonry walls, 104
 mineral wood, 102
 movements, 106
 platform construction, 111
 termites, 111
 vapor barriers, 105
Wood classification, 18
 defects, 19
 dimensional stability, 21
 durability, 20
 fire resistance, 21
 glued-laminated wood, 22
 hardwoods, 18
 insulating qualities, 21
 plywood, 22
 seasoning, 19
 softwoods, 18
 strength, 20
 unit weight, 21
 wood composites, 21